STAYING HEALTHY
AT WORK

Staying Healthy at Work

Lifeskills International Ltd

Lifeskills team of writers:
Barrie Hopson
Jack Loughary
Steve Murgatroyd
Teresa Ripley
Mike Scally
Don Simpson

Lifeskills Series Editor: Jonathan Kitching

Gower

Published by
Gower Publishing Limited
Gower House
Croft Road
Aldershot
Hampshire GU11 3HR

Gower
Old Post Road
Brookfield
Vermont 05036
USA

Lifeskills International Ltd have asserted their rights under the Copyright, Designs and Patents Act 1988 to be identified as the proprietors of this work.

British Library Cataloguing in Publication Data
Staying healthy at work
    1. Industrial hygiene 2. Job stress 3. Employees - Health and hygiene
    I. Lifeskills International Ltd.
    613.6′2

    ISBN 0 566 08144 X

Typeset in Middlesex by Setpoint and printed in Great Britain at the University Press, Cambridge.

CONTENTS

1

THE SLIPPERY SLOPE OF STAYING HEALTHY

Purpose

This book has four basic purposes:

◆ to assist you, as a manager, in becoming continuously aware of the status of your health
◆ to help you learn several health concepts and techniques and plan a staying healthy routine
◆ to demonstrate ways of following through your staying healthy routine
◆ to demonstrate how staying healthy is the cornerstone on which other life roles, and probably satisfactions, depend.

Here are the topics covered in this chapter. First, a working definition of health is provided. Second, the notion of health style is defined. Third, there is a discussion about health awareness, including an invitation to make a brief personal health-awareness assessment. Fourth, the concept of mind and body health is introduced.

Health defined

What do we mean by 'my health'? To get you thinking, how would you define health?

What is your definition of health?

_____

_____

What, in your opinion, might be a difference between physical health and mental health?

If we cut right to the heart of health and consult the dictionary, we find the following:

Health: 1. physical and mental well-being; soundness; freedom from defect, pain, or disease; normality of mental and physical functions ... Health is a universal good condition.

Healthiness is ... the state or condition of being healthy.

How do your definitions compare with that?

Health is a preferred condition of mind and body. It is also thinking, behaviour and attitudes aimed at maintaining that preferred condition. True, everyone alive has some sort of health. It may be poor or nearly gone, but given life there is hope for health. For all practical purposes, however, health as used here refers to the positive or desired end of the continuum of physical and mental well-being.

Operationally speaking, health means that the several physical body systems are functioning well. You are able to breathe, eat, digest, eliminate waste, move your arms and legs and, if appropriate to your age, reproduce. You may not be able to walk, run and jump, nor see, hear or speak perfectly due to age-related causes or accidental impairment, yet still have good health. Health is relative to your time and conditions in life. It is not the same as physical strength, although it is related to it. An Olympic athlete may be unhealthy in some ways and a paraplegic have no health problems. The first may be physically fit but use drugs and be depressed; the latter may be physically immobile but a model of healthy behaviour. Health is being reasonable about your mental and physical self most of the time.

Health is relative: you need to be aware of what it means to you and work at maintaining and improving it. Some things are out of our hands, but awareness is not one of them.

Health style

If you have attended personal development training sessions or read self-help books, the chances are that you have considered or been introduced to the concept of lifestyles. There are many ways to view lifestyles, each with its own focus. These include financial, travel, social, work, relationships, recreation, diet, fitness ... the list goes on. Each lifestyle focus usually contains a number of different levels and comparing one's lifestyle with those of one's peers can be an interesting exercise, or at least some find it so.

Regardless of the lifestyle you may have chosen or drifted into, a crucial underlying consideration is your health style – crucial because whatever the main focus of your lifestyle, from jetting around the world to spending weekends doing good works for others, your satisfaction will depend to a large extent on your good health.

Why, then, does the obvious value of staying healthy often get lost in the rush for life satisfaction? One reason is that it is a broad, general idea. Like patriotism or saving for retirement, it is such a good idea that we often take it for granted; until, of course, a health crisis occurs. When that has been resolved, we become serious about taking specific steps for staying healthy – for a couple of days, at least! Then other concerns become more interesting or seemingly demanding (a matter of perception) and health awareness goes the way of many tedious resolutions ... until the next crisis occurs. Incidentally, these crises arise at shorter intervals as we travel through life.

Health hint
Awareness is the first step of the staying healthy triad of awareness–action–awareness. That is, be aware of health conditions – act on your awareness – be aware of the impact of your action. Let your experience assist you in staying healthy.

Health awareness

How can we maintain health awareness? How, in other words, can we track and attend to the specifics? This may be second nature to you, but that is not a realistic expectation for most of us. Another way to maintain health awareness is to view it as involving a set of skills that can be learned. Many concepts underlie such skills and these will be discussed in due course. Let us move to an illustration and then a personal application.

Imagine, for example, that as part of your work as an assistant manager you are assigned the task of making a 30-minute presentation to senior managers in your organization. The topic is: 'A view from within: the organization's strengths and weaknesses'. Careful person that you are, you spend a considerable amount of time talking with other managers and colleagues about their views and thoughts. When you are satisfied that you have adequate information for your presentation, you prepare an outline and ask two other managers to review your presentation. You consider their suggestions, refine your presentation and practise it several times. Finally, you think that your presentation is as clear as you can make it. You leave work on Friday knowing that you are well prepared.

You think about the presentation briefly on Saturday and on Sunday evening you review it briefly. Being pleased with your preparation, you decide to sip a glass of wine or two and enjoy the late film on television. Plenty of time to do that and still get a good night's sleep and be ready to make the presentation in the morning. Except that after about an hour of sleep you wake up. Tummy a bit bubbly, you get up and head for the loo. Feeling relieved, you return to bed, anticipating the open arms of Morpheus. Surprise! Morpheus has gone, replaced by a vision of the auditorium full of senior managers waiting for your presentation. After a few minutes of tossing and turning you remember one of your private great truths: challenge is often accompanied by stress, which in turn successfully spoils a good night's sleep. After what seems to be a 36-hour night of wrestling the sheets and pillows, your alarm clock finally goes off and, though tired, with some degree of relief you get out of bed and start your morning routine. Before long, there you are, entering the presentation room, tired, wondering if your presentation will be well received and feeling more than a little nervous.

If you can identify at all with this scenario, we can use it to make a couple of points regarding the importance of health awareness:

◆ What could you do at this very belated state of health awareness?
◆ What could you have done, in the cause of health awareness, to avoid the Monday morning disaster?

What could you do at this very belated state of health awareness?

Being aware of yourself at the moment is your best tack, but it is difficult to achieve because you are too busy attending to the past or future. You might be thinking of past times when a speaking assignment went poorly. Or you may be preoccupied with the future, wondering what you will do if things start going badly in the middle of the presentation. In other words, your attention is diverted to thoughts that really have nothing constructive to do with delivering an effective presentation. But don't worry! It is never too late to become aware and focus on the existing issue. That is, you can turn your attention to the specifics of your current problem, which is not so much the outcome of the speech, but the giving of it.

There is a mental health skill called 'self-talk' that might serve you well at this point. Self-talk is pretty much what you probably did as a child – that is, hold a conversation with yourself in your head, both asking and answering the questions. Adult self-talk has an important survival purpose: asking the questions that are of most concern and providing reassuring answers. Your self-talk might sound like this:

Worry: 'I'm nervous. I'll never be able to get through this without a lot of mistakes. I'm afraid of failing and making a laughing stock of myself.'

Answer:'It is natural to be nervous. Only a fool would not be. It helps to get the adrenaline flowing. Certainly, I may not be as polished as I want to be. But I'll notice that more than the audience. Most of them understand that I am new to this and will be supportive. I would be if I were in their shoes. Failure is relative. I'm prepared and have the contents of the speech well in hand. Half the battle is over. I'll find a friendly face and focus on that person as my key audience.'

Worry:'What can I do if I do begin to feel really uncomfortable?'
Answer:'Slow down. Take plenty of breaths. Smile. Smile again.'

Maybe you will still feel queasy, or perhaps not. Either way, you are now aware of some specific concerns and have thought them through. That is a much better place to be than nervously fretting at the podium, not having thought through and responded to your concerns.

What could you have done, in the cause of health awareness, to avoid the Monday morning disaster?

Now consider the situation if your health-awareness sensors had been turned on ahead of the event. It would have occurred to you during the fatal evening that you were beginning to feel somewhat stressed about the next day's assignment. There were the usual signs: you were fidgety, couldn't settle on an interesting task, gulped down the evening meal, and were more or less preoccupied with continued mini-planning of Monday's presentation. If you had been in a state of health awareness, the sequence of events, beginning about Sunday afternoon, might have gone something like this.

As is your habit, you take a few minutes on Sunday afternoon to think about the next week's work. Aha ... Monday is the Big Day, you recall. And what do you know about Big Days? You know that they have been known to precipitate minor stomach upsets and sleeping difficulties. What do you do when you become aware of such an impending situation? Experience has taught that it would be wise to plan a health strategy. For example, you could find an especially involving project for

Sunday afternoon (you could play golf or take a relaxing walk in beautiful countryside). That helps get your mind off the Big Day. Then, silly as it may seem, you write yourself a note about eating slowly and not combining dinner with some distracting activity. Last, a hot bath and a short read seem to help prepare you for a good rest and so that is what you decide to do. You take a warm bath, then pick up that good novel or magazine where you left it. Then, as sleepiness draws in, you crawl into bed, relaxed with a mind and body ready for sleep.

Granted, you may not like hot baths in the evening or novels before nodding off, so substitute whatever might work for you! You don't know what works for you? If you were to think about it, you could probably come up with some ideas and develop such a routine, couldn't you? Don't let it go unnoticed that the key step in this adventure of health awareness took place (or should have) on Sunday afternoon when you devoted a few minutes to previewing the coming Monday. It was then that your routine made you aware (as in health awareness) of a critical event on the horizon and, equally importantly, led you to make a decision to do something healthy in preparation for it.

However, don't forget the first point: no one is perfect and awareness of health issues may be late arriving. Even then, you can be prepared with a small reserve of stress-management techniques.

Health hint
Being aware of a health problem is often 50 per cent of the solution.

Personal example
It usually makes topics such as health awareness more interesting if we can establish a personal connection. Here is your opportunity to personalize the health-awareness idea.

Thoughts–feelings–body-awareness technique
To begin, use the space overleaf to describe a past situation in which you experienced unpleasant health feelings, thoughts or other symptoms. Be specific.

Describe your thoughts as you recall them. Be specific.

Describe your feelings as you recall them. Be specific.

Describe the symptoms that your body was experiencing. Be specific.

Congratulations! You just demonstrated the thoughts–feelings-body-awareness technique. Being aware of a stressful situation by describing concerns in terms of thoughts, feelings and body awareness can be 50 per cent of the solution!

The mind/body health concept

There is increasing understanding of how thinking contributes to illness. People have had strong suspicions about that for ages, but it is nice to have scientists agree. To be more accurate, it is only recently that scientists have been able to confirm that mind/body associations exist and to clarify how they act.

One obvious example is via a bio-feedback machine that is set to measure heart-beat rate. An electric wire from a computer is attached to your wrist. As it senses your heart beat, the computer displays each beat as a bleep on its monitor. It is not difficult quickly to teach people to change the rate of their heart beat simply by thinking about it, or about some particular event. Peaceful thoughts relax and slow the heart beat; frightening thoughts increase the rate. As you observe the results of your thinking on the screen in the form of a change in the rate of bleeps, you learn how to control your heart-beat rate. Soon, you can control the rate simply by being purposeful about your thoughts and without the aid of the computer.

There are a variety of practical applications of this mind/body relationship. Among these are reducing fear about such activities as flying, making presentations, dealing with conflict, uncertainty, meeting new people and many other situations that cause anxiety. The significance to staying healthy of such advances in understanding of the mind/body phenomena is that we can learn to influence, if not control, an increasing number of health conditions. For example:

◆ It is now clear that we can learn certain eating behaviours that enable us to control our intake of food. That in turn can allow us to influence such conditions as blood pressure, cholesterol levels and other chemicals present in the body.

◆ One researcher reported that different coping strategies can be associated with changes in emotional stress (Bolger, 1990).

◆ Other researchers suggest that religious orientation and belonging to various faiths and denominations plays an important role in managing life stress (Park, Cohen and Herb, 1990).

◆ Several researchers suggest that a psychological condition called 'hardiness', known to help in coping with situations that many people find stressful, can be learned (Florian, Mikulincer and Taubman, 1995).

How can the mind/body connection be summarized? Quite simply, it is that your state of mind (for example, thoughts, feelings, moods and images) can affect your body's systems and thus your health. You can't wish yourself to perfect health, but it has been shown that:

◆ anger and hostility contribute to heart disease, and you can learn to deflate both (Williams, 1993)
◆ people without a social network are more likely to become ill, and you can learn to develop a network of friends and associates (Spiegel, 1993).

Similarly, it is well established that when people are adjusting to major life changes, their health and performance can suffer. The more changes, the greater the stress that people seem to experience, but again, it is possible to regulate many such changes.

In addition, there are methods for using our minds to:

◆ reduce the side effects of chemotherapy (Holland and Lewis, 1993)
◆ cope with migraine headaches (Turk and Nash, 1993)
◆ help deal with chronic pain (Turk and Nash, 1993).

People who learn to be aware of their mental states and how to influence them can help themselves remain healthy and deal with illness when it occurs – a place we would all like to be. This is called self-help. The price is right and it is readily available, too!

Self-help techniques

In this book many self-help techniques will be described. They are an opportunity for you to understand further the techniques available and to use them yourself. This chapter ends with two techniques, one using your body and the other using your mind. The effect of both can be increased awareness and stress reduction.

Taking a breath break technique

You've heard of coffee breaks and cigarette breaks. One puts a stimulant in your body and the other pollutes your lungs. Be kind to yourself and start a new routine – take a breath break. The benefits are many. It can decrease blood pressure, reduce anxiety, improve energy, improve digestion and reduce stress. The risks are nil (provided you are not in a polluted airspace), the cost is zero and your breath is always with you. No special equipment is needed.

You may think this sounds boring. Newcomers to the art of breathing are asked to imagine putting their head in a bucket of water for a minute. Very quickly breathing goes from boring to essential. And indeed it is! You can live for some time without eating and drinking, but not without breathing. It is the one function of the body that is both involuntary and voluntary. It is a tool you can use in staying healthy and assisting your body to be at its best.

Are you interested in trying the breath break technique? For this week only, take a five-minute breath break each day at work. If you don't have five minutes, try three. Anyone can find three minutes in a day.

Sit at your desk or someplace where you will not be disturbed for five minutes (the toilet, for example). Once settled, close your eyes, focus on your breathing and breathe normally. If you begin having thoughts, gently stop them by focusing back on your breathing. To do this, say silently to yourself 'in' each time you inhale and 'out' each time you exhale. Continue repeating this until you are able to focus solely on your breathing. This sounds deceptively simple to do, but can be difficult to achieve. With practice it will get easier.

Open your eyes after five minutes. How do you feel? Probably better than when you started.

Take a five-minute breath break each day.

Thoughts–feelings–body-awareness technique

Previously in this chapter you were asked to review a past stressful situation in terms of the thoughts, feelings and physical signs that you experienced. Now let's consider your present life with the goal of making you more aware of stressful situations and how you react (see Table 1.1).

Summary

Health awareness can be developed and maintained. Health is such a general and broad concept that in order to take care of ourselves it is important to become aware of the specific aspects of health that are involved. Having health-awareness strategies provides an excellent basis for making important health decisions and using health techniques.

Application

One suggestion for adapting the material to your own health concerns is to review this chapter, making a list of the techniques described. Then highlight those items on your list that are especially important to you and incorporate them into your diary or planner for the next four weeks. As you try each technique, take a moment to evaluate its usefulness to you and adjust the following week's schedule accordingly. You may even find that you make significant modifications to the techniques we have described and, in this way, tailor the exercises to suit you and your lifestyle better. What works best for you, and your health, is what is most important.

Table 1.1 *Thoughts–feelings–body-awareness technique*

Make several copies of this page and keep some at your home and your office. Complete the form during or after several situations that you find stressful.

What was the incident? Be specific.

Describe your thoughts. Be specific.

Describe your feelings. Be specific.

Decribe your bodily or physical signs.

At the end of the week, review the forms that you have completed. What are you aware of as far as your thoughts, feelings and body signs are concerned? Record that information below. Remember that awareness is half the battle.

What does your review say about you and your sources of stress?

What does it say about your ways of coping at present?

What do you think you need to work on?

2

A SHORT HISTORY OF STRESS

Cave people you may recognize

Let's assume that you come from a long line of managers and, if we can go back far enough, we can pick up your family history with Ugg. Ugg, your managing ancestor, supervises a group of people whose job is to separate big rocks from not so big rocks. There is a purpose to their work, in that only big rocks can be used to dam streams. Stream Damming Ltd pays Ugg well for getting the most out of the rock sorters. Each rock sorter is paid one fish for each satisfactory day of work. Satisfactory is defined as getting no more than two rocks in the wrong pile.

The main source of stress in Ugg's job is that some managing director has defined a big rock as one too heavy for a worker to move or, to use the technical term of the age, shove. When a rock sorter finds a rock too heavy for him or her to shove, it is marked as big and later a crew of rock movers comes along and moves it into the river. Simple, isn't it? Not always.

As you would imagine, the rock sorters come in differing sizes and strengths. Thus it is not unusual for less than strong rock sorters to mark the big sign on a rock when it is in fact not a big rock. How is this discovered, you may ask? The answer is that all of the rock sorter crew bosses are big, strong people, having survived through the ranks over the years. They test shove each rock marked with the big sign and when one moves the responsible rock sorter is in big trouble! A mis-sorting error has been made.

One day Ugg is reading his paperwork and discovers that Nog, one of his smallest rock sorters, has a sorting error rate that exceeds the maximum. So Ugg refers to the Stream Damming Ltd policy manual to determine the appropriate penalty. 'Dock the worker one half of a fish,' reads the manual. Consequently, just before finishing time, when the workers are about to line up for their pay, Ugg slices a fish in half and gives it to Nog. Nog takes one look at the half fish, grabs it by the tail and begins to hit Ugg in the face with it. A couple of co-workers pull Nog off Ugg, so no physical damage is done. Nevertheless, Ugg still has a very stressed employee on his hands. 'What in the world was egging Nog on?' he wonders. Several possibilities occur to him. What do you think?

Egging Nog questionnaire

Rate the following possible causes of stress from highest to lowest (10 = high).

1 This is the last straw (or rock) for Nog, following a series of unfair performance appraisals.
2 Nog is primarily upset over the obtuse criteria used to measure rocks.
3 Nog's wife is counting on a whole fish because they are having company for dinner.
4 When Nog was a child his brother cut his pet fish in half and Nog still gets upset over cut fish.
5 Nog prefers the fish head and is mad because he got the fish tail.

If you were Nog, what would you find stressful about the situation or incident?

_____

_____

What is really causing Nog's stress? It turns out that he had been misinformed about the symbolic meaning of a half fish during the employee orientation workshop. He thought that it signified the loss of his job, not of half a day's pay. When Nog saw the half fish, fear set in. He immediately worried about how he would support his family, became stressed and directed his frustration at the nearest authority figure in sight, manager Ugg.

So much for Nog's stress reaction. What about manager Ugg? It doesn't take much experience to see that he was suffering his own stress reaction when Nog set about him with a fish half! Research on stress tells us that historically there were two basic reactions to it. These, as you may know if you were paying attention during your last stress workshop, are fight and flight. Ugg, scared and thus feeling a great deal of stress about what Nog might do to him, had two means of coping with his stress, according to the research. He could jump out of the cave and run down the path, thus using the flight response. Or he could grab his club and wallop Nog in the head, thus using the fight response. However (and this is important), Ugg had been to Manager Sensitivity Training School and had learned one, that managers get fired for running away from problems and, two, that whacking employees on the head with a club is no longer an acceptable mentoring technique at Stream Damming Ltd. Thus here was your ancestor Ugg the manager, stuck between a rock and hard place, without a socially acceptable outlet for his stress. That could be a very serious condition for both management and staff.

Fortunately, nature took over. Ugg was so stressed that he peed in his deer skin, which relieved everyone's tension and all had a good laugh. A couple of the other workers felt sorry for Nog and threw him a whole fish for his wife's dinner party, and Ugg later said that he would work on getting the performance measure changed.

Thus, as you can see, stress was neither all bad nor all good in the olden days.

Stress and high technology

What is your assessment of stress in today's world? Has the magic of technology led to less stress at work, home, play and in other aspects of society? What do you think?

Stress and technology questionnaire
What are the main effects that technology has had on your stress level? List those that stand out, both positive and nega-

tive, and mark each plus or minus, or both, according to your opinion.

If you could make significant changes to your working life and environment aimed at making it healthier for you, what would they be?

Now consider your non-working life. What are the main effects that technology has had on your stress level there? List those that stand out, both positive and negative, and mark each plus or minus or both, according to your opinion.

What changes would you make in your non-working life aimed at making it healthier for you?

Take a few moments and think carefully about your answers. Then briefly note what stands in the way of making these changes. The question is not intended to catch you out. It is one to which many people have given little consideration. What is your response to it?

It would be more meaningful if we could read your reactions and respond to them directly. However, based on the responses that others have made to similar questions, it is possible to make some educated guesses regarding your comments.

First, people's complaints about stress at work and outside work vary tremendously. You've undoubtedly observed that yourself. Sound transmitted on radio or television may be perceived as music (both soothing and exciting) by some, while others perceive it as noise (both distracting and maddening). The constant attention of a supervisor may be reassuring to Susan but produce stress for Sam. A carefully detailed schedule of the day's tasks contributes to an organized and thus manageable and calmer working day for careful Carl. The same clear and precise schedule inhibits the productivity of creative Colleen and, for her, results in frustrating and non-productive stress.

And then there are those bothersome questions: 'Can stress be productive? How can stress be both debilitating and productive?' The answer is simple and depends on the individual, as the previous paragraph implied. There is no doubt that a certain amount of stress can be motivating. Without it, some people would never get out of bed and, without it, a lot more people would not achieve anywhere near their capacity, which may or may not be a good thing. Ever thought about working in an environment where everyone 'achieved their capacity'? Yuk!

However, practically speaking and depending on an individual's personal make-up, stress can provide positive and productive motivation. Once it exceeds an individual's ability to use it effectively, stress can be, and usually is, destructive. More about this later on. For now, let's agree that we are talking about stress that exceeds your (or the next person's) ability to use it effectively.

The second generalization concerning the comments that people make about stress is that they are mostly straightforward and actionable. One gets the exceptional hanky waver or space cadet, but by far the majority of people seem reasonable about their desires concerning stress levels and sources of stress, both at work and away from it. For example, people wish that they didn't have to commute so far or under such unpleasant conditions; feel that policy and procedures change so fast that it is impossible to keep up with them; think that management (meaning the next level up) doesn't understand the details of work in the trenches or, conversely, doesn't see the really big picture. People also note that they would like more work space, to have a greater voice in the decision-making process at work, to set more of their own priorities, to receive more rewards for their achievements and not to be so rushed to get things done on time. These are reasonable and widely experienced stress concerns.

Comments about stress in non-work situations run along parallel themes. Life is too rushed, there is too much to do and not enough time to do it in and the pressures from family and friends take much of the joy out of non-work activities. There are also concerns regarding not enough variety in life, too few friends, an uncooperative and unreasonable family, the shortage of close friends and insufficient financial resources.

Both of these kinds of observation lead to a third generalization regarding stress in our high-tech society. The core of complaints and concerns about stress in a high-technology society are much more about society than about high technology itself. Think about that and about your own responses to the questions that you answered at the start of this section. Were they about the nature of electronic machines, the technology of data

transmission, satellites wandering out of orbit, conflicting engi-
neering designs or even the malfunctioning of equipment?
Probably not. More than likely the emphasis was on people and
human interaction.

Managing stress

Ask not what is stressful about your computer, but about the
stressful conditions that surround its use. Stress in a high-
technology society is not so much about the technology itself,
but more about the people and circumstances associated with it.
The value in looking at stress from this perspective is that you
can have a considerable impact on how you manage it, regardless of your technological expertise. Here is a summary of
options suggested by many writers on the topic of managing
stress.

Leave the situation that is stressful for you
That is a pretty desperate move, but worth keeping on your 'list
of possible stress-management strategies', for a while at least. In
a work situation, this often translates into resigning, obtaining a
horizontal move such as changing departments, or asking for a
demotion. The latter may be viewed with suspicion by those
'higher up', but it has enabled people to get out of stressful job
situations without losing their incomes. The feasibility of leaving as a means of dealing with a stressful situation depends on
many variables. Two or three of the most important are your personal resources, the kind of work you do and your ability to deal
with the ambiguity often involved in job change. Among
important personal resources are funds to hold you over if a total
job change is involved, the psychological support of your partner or family, if you have one, and professional connections
both within and outside your present organization.

The ease of changing positions depends largely on how readily you can take your skills and connections with you. If you are
in one of the traditional professions, the opportunities for
changing positions is probably greater than if you are in managerial work. However, given changes in business and other

working environments, people in areas such as computing, writing, marketing, sales, finance and other formerly 'company-based' jobs can often easily move from one position to another, depending on their connections, reputation, specific competencies and, of course, the market demand.

The ambiguity and potential insecurity involved in such dramatic changes in work represent a serious kind of stress in themselves for many people. Only you can judge, but the important question is: 'Will the change be more stressful than staying with the present situation?'

Leaving stressful non-work situations may also be worth considering as a stress-management strategy, but this may be even more difficult to make successful. Marital separation, divorce, moving home and other such domestic decisions are more quickly made than implemented. As many people have learned, it may take a long time to obtain a divorce. The conditions noted regarding changes in work situations also apply to dramatic non-work changes. These options require careful consideration before taking action.

Stay and accept the situation

This stress-coping strategy appears on the surface to be one of the most frequently used. But is it what it appears to be? All you need to do is look around your workplace to see people who claim to have analysed their dissatisfying work situation, recognized the costs of leaving it then decided to remain ... and then began devoting a good part of their effort to travelling down the path of resentment to some full-blown illness. A reasoned, thoughtful and planned acceptance of the status quo still need not be a poor decision, so long as it really is reasoned, thoughtful and planned. It can be a viable means of managing stress with manageable side effects. Techniques for how to do that are illustrated in Chapters 3 to 6.

Change the situation

This approach to stress management refers to adjusting 'components' of work and non-work situations. It is a strategy aimed at lessening particular stressor points while leaving the general situation essentially intact. For example, a physical work space can be modified. You have probably seen people hang pictures or posters to create a softer ambience in a work space, or done it yourself. One group of workers established a pool of moderately expensive wall hangings that each person could borrow for a set time. Another person, assigned to a windowless cavern miles from an outside opening, created seaside murals on two walls by carefully arranging inexpensive but attractive photo posters. Moving furniture to create more private and comfortable space is another means of managing stress through changing the physical environment. Basic to this approach is analysing a situation, identifying the particular physical stressors and being creative about stress-reducing solutions.

The stressors in a situation can also be a matter of people and their physical and temporal relationship to one another. A flexible work schedule, where people share common working hours but are accorded degrees of discretion regarding the pattern of their working day, is an example of managing stress through allowing personal time discretion. Time restrictions and confinement can also be stressful. Allowing workers who are routinely restricted to work stations a reasonable amount of time for breaks and personal phone calls could reduce stress for some people.

These, and similar stress-management solutions, may be provided by employers as a matter of policy or can be formulated on an individual basis. Much depends on the nature of the work and what a supervisor or employer may get in return for requested adjustments. 'Because it has never been done here before' is a questionable reason for not requesting or trying a particular strategy. An enlightened employer will provide means for workers to make such requests without fear of comeback and for open discussion of the implications of making them. Even though this is not a book about labour relations, the implications for managing stress by changing work situations are well worth considering.

Change yourself
Skills may be purely technical and job related, or more general, such as the ability to relate well to other people. If stress appears

to be, in part, a result of lack of skills and competencies, then changing yourself – that is, adding skills and competencies to your 'behaviour repertoire' – is another stress-management technique worth considering. In a more formal sense, this could mean completing a training course or seminar in order to acquire competencies ranging from assertiveness to word-processing skills. Sources of personal change courses and other training programmes are employers, local education institutions, professional associations and the commercial sector.

In addition to utilizing traditionally structured courses, personal change may be accomplished by participation in a wide variety of ongoing groups. Church and recreational organizations sponsor such activities. For example, if stress seems to be generated by your difficulty in engaging in 'small talk', then one option is to join a group in which such informal verbal exchange is the name of the game. Such groups allow you to observe and try various styles of communicating without the stress-producing conditions of the workplace. It may seem somewhat contrived, but think seriously about this source of personal change before rejecting it out of hand. Once developed, such skills will probably transfer easily to a job context and serve to ward off at least one kind of stressor.

The underlying point is that no one need find the status quo inevitable. If you identify stressors in your life and they appear to be related to current behaviours or to a lack of skills, then changing yourself is a reasonable solution. In this regard, it may be useful to think of the notion of 'minimal change'. The idea is simple: a strategic small change can lead to larger differences.

Imagine a pie-shaped chart, with the outer rim being the status quo. Picture a triangular piece of the pie. Changes made at the outer rim are considerably larger than those at the centre. A small change at the inner tip can easily grow to a larger change at the outer rim. Corny, perhaps, but think about it. For example, smile to yourself five times in a mirror before you leave for work and it will be easier to smile once at someone on the train. Smile at someone on the train and it will be easier to smile at someone at work. Begin smiling more at colleagues and some of them will be more friendly to you. (A few may think your behaviour a little strange, but that may be interesting too!)

Change your thinking, feelings and behaviour

So much for the easy stuff. It is also possible to manage stress by changing how you think, feel and behave in specific situations. However, the techniques needed to achieve the changes generally require more practice and commitment than the techniques referred to so far. Chapters 3 to 5 describe a variety of cases, each demonstrating the use of at least two stress-management techniques.

Each case reflects the awareness–action–awareness triad presented in Chapter 1. Remember that all stress management begins with an awareness that something is not as you would like it to be. Your awareness may be far from satisfactory or complete, but the irritation, and thus motivation for fixing the condition, has reared its head. Action, the second step, may be essentially random or by chance to begin with, but it is probably based on some hunch you have about the cause of your discomfort. The action may be far from satisfactory, but your observations of what did or did not happen as a result of it should provide useful information for refining your next action. And so the awareness–action–awareness cycle continues until the discrepancy between the way you want things to be and the way you perceive them to be is tolerable to you; that is, until your stress is being managed to your satisfaction. The cycle, as you will have realized, is not unique to stress management and it is more than trial and error by guesswork. It is trial and error by informed action and observation. The latter, as you probably recognize, is called feedback.

It's up to you

If this approach to stress management appeals to you, then it is time to move on to the next four chapters. Here are two approaches that you might take. The first is to read straight through the book to get an overview of the techniques. Then select those that fit your needs and interests and attend to them more seriously. Many people like to get the big picture first and then move to the specifics.

The second approach is to read the first case in Chapter 3. This will acquaint you with the format used. You may or may not be particularly interested in the problem that it addresses. Notice that each case has an annotated title, so you can quickly skim the titles and choose the cases that appeal especially to you. Then it is a matter of mixing and matching your interests with the case scenarios. The three chapter divisions, 'Body', 'Mind' and 'Behaviour', tend to stand even if there is still overlap between different kinds of problem and the procedures and techniques described. There are no perfect matches, but then life is like that.

Chapter 3 is a collection of additional stress-management techniques and procedures designed to enhance mental health. These are presented in a self-explanatory format.

Connecting

Before moving on to Chapter 3, it may be useful to think again about the awareness–action–awareness triad discussed in Chapter 1; that is, be aware of health conditions, act on your awareness and be aware of the impact of your actions. Then, reread the thoughts that you have recorded in this chapter and consider them in terms of these three elements. You may notice how using the triad could have made a difference in your own circumstances.

View the technique as a personal feedback system that helps you expand your 'personal awareness information bank'. That growing information bank is, in turn, a resource that can assist you in taking increasingly effective actions. This is not a new idea; it is simply one explanation of the nature of learning from experience. The point is that our greater understanding of the role that awareness plays in our behaviour can enhance the learning.

3

THE BODIES IN QUESTION

Introduction

The cases in this and the following two chapters begin with complaints that people have about their health. This chapter concentrates on complaints that are primarily physical in nature – people complaining about some bodily ache or pain. As discussed in Chapter 2, the distinction between body and mind in health appears to be less clear cut than we may think. There is also considerable overlap between what we do to maintain physical and mental health respectively. The cases discussed in the book attempt to illustrate that concept.

Before examining the cases, it is important that we establish some basic ground rules. The first is that there are no magic cures to health problems. A magic cure is one that has no scientific foundation. There are, it must be acknowledged, some non-scientific based cures and recommendations that may seem to have desirable outcomes. When and if they are researched, they may or may not prove to have validity. The point to be made is that this book is not concerned with pop health cures, fads or magical prescriptions.

The second ground rule is that staying healthy should begin with your being aware of the current status of your physical health. How can that be achieved? Probably the most feasible means is a physical check-up and discussion with your doctor or health consultant. In other words, our recommendations regarding the so-called self-help techniques discussed here are made with the assumption that you have consulted a health expert about abnormal concerns and that in addition you have acquired an evaluation of your present standard of health.

The third ground rule is that if in the course of reading the book you become aware that your condition or behaviour is not within the range of good health, you take steps to become aware about its significance. Examples of such conditions and behaviours include excessive consumption of food, a diet inconsistent with standard recommendations, excessive alcohol and drug use, smoking and a lack of physical exercise. By their very nature, abnormal mental and emotional conditions and behaviour may not be as easily recognizable, but personal observations and interactions with others may identify reasons for seeking professional advice. These would include any uncontrolled emotions, desires and behaviours.

Having clarified the ground rules, let it be said once more that the techniques and ideas discussed and recommended in the cases all have reasonable scientifically based foundations. Given the above ground rules, they can be employed without professional supervision.

Bert: a case of perfectionism

Complaints

Bert is a 34-year-old personnel manager. His primary complaint, at least the one he would verbalize most often, is that his stomach is frequently upset. He belches several times an hour and, when he feels especially pressured or tense, he feels he has stomach indigestion. In his words, he has abdominal discomfort and heartburn. That is often accompanied by constipation. Bert has two children, both in fee-paying schools, and has a mortgage on his house. The latter is one reason that his wife, Susan, decided to return to work part-time as an infant school teacher. Bert's view of himself is that his marriage is successful and happy and he is optimistic about his career development. He would like to become head of a human resource company or major division in the next four or five years.

Bert has an admirable career history. After leaving school, he began as an insurance claims representative in a small town. He was a high-volume producer, often working longer hours than required. He took leadership roles both in professional associations and his local residents' association. Noticed by his company, he was rewarded with a promotion to its head office in the city. This entailed selling one home, buying another and re-establishing himself and his family in a commuter suburb. Bert's wife has described him as an overachiever and a workaholic. If pushed, she goes on to say that Bert is bright enough, but he seems to aspire to higher levels than appear to come naturally. She also thinks that his continuing to work many evenings and Saturdays is not normal, given the high level of success he has obtained with his company.

Bert first consulted a doctor about his 'stomach problem' around five years ago. After a physical examination, he was told that the problem was probably caused by excess stomach acidity. The doctor prescribed an antacid drug and suggested that Bert try to avoid fatty and spicy foods. Bert complied, but the stomach complaints continued. Six months ago he was examined by another doctor, who confirmed the earlier diagnosis. He prescribed an alternative antacid drug with the assurance that if Bert complied with his prescription the stomach symptoms would be reduced. The doctor also recommended to Bert that he 'take it easy' on weekends and pursue a hobby or activity that would help him get his mind off his work.

Bert has complied with the drug prescription, but has not initiated leisure activities. He did discuss his preoccupation with work with Susan, who agreed that he had a problem. She added that she had just about given up on his workaholic nature and had little hope that he would ever change. Bert's first reaction to this discussion with his wife was anger and frustration. Those feelings soon diminished. However, he now finds himself preoccupied at times over what he perceives as his wife's negative evaluation of him. His stomach discomfort continues and he now complains about tension in his shoulder and back muscles. These are distracting to the extent that he is finding it difficult to sit at his desk for longer than half an hour.

Discussion
You probably think that any fool could see what is bothering Bert and you are probably right. However, it is not unusual to be so involved in our own concerns and behaviour that we fail to make common sense out of them. It is probably accurate to say

that whether or not he is aware of it, Bert's preoccupation with high performance at work is related to his several health issues. We do know that psychological stress can lead to physical tension and pain. Assuming that is an idea that Bert would consider, then what self-help remedies might be suitable?

Bert could try to clarify his life goals as a means of setting priorities. One question he might pursue, for example, is whether it is exceptional performance he is after or recognition from his managers that he works as hard as he does. If he suspects that it may be more of the latter than he previously thought, then he might think about whether his managers would actually be concerned with his performance, as long as it were acceptable and valued. Perfectionism is often mostly in the eye of the doer. To believe that others are equally concerned with our achievement is to hold on to a somewhat adolescent notion – one that can do little about achieving perfection, other than ending up with frustration and disappointment and, in Bert's case, a lot of unrewarded effort. In addition, if he took time to reflect more on his situation, Bert might explore his working life in juxtaposition with his family and personal concerns. Perhaps he might then see that his wife has been asking for greater participation in family matters.

There are procedures designed to assist people in clarifying values, setting priorities and making decisions. These are described in other cases. Remedies aimed more directly at physical and cognitive complaints such as Bert's are described in the next section.

Strategy

We now jump ahead to a point where Bert is ready to implement self-help techniques for reducing the stress that we assume is related to his physical symptoms. Many such techniques have been described in both lay and professional literature. Some techniques are designed to assist in body relaxation. Learning to relax is a skill and, like any skill, it takes practice. Two relaxation techniques are described next. Bert or anyone else contemplating using these would be asked to read the instructions and assess whether or not he could incorporate these in his life for the next two weeks. Two weeks is important, because it will take that long to see if a technique is having the desired effect.

In addition to benefiting from relaxation exercises, it seems clear that Bert might gain some insight from learning more about thoughts that influence what some might call his workaholic behaviour. One explanation for such behaviour is identified in the concept of 'shoulds'. Most people recognize that over the course of our lives we pick up beliefs about what we should and should not do. Even though many such beliefs are useful guides for how to behave, some of the 'shoulds' that shape our thinking and behaviour may be irrational. A cognitive exercise for clarifying 'shoulds' is explained shortly and can be completed in less time than the relaxation technique.

The relaxation response

The relaxation response was first described by Dr Herbert Benson in 1975 (outlined in Benson, 1996). When a person is under stress, as described in Chapter 2, the body responds to the threat or perceived threat by preparing for fight or flight. Metabolism, blood pressure, heart rate, rate of breathing, blood flow to the extremities and muscle tension all increase. The amount of increase depends on the perceived threat. The only decrease is found in slower brain waves.

Benson discovered that by having people practise the relaxation response the body could, as he termed it, downshift and not have to work so hard. The breathing slows, the heart doesn't have to pump as many times and, in general, the body can have as real a rest as it does during a good night's sleep.

Since the body spends much of its day ready for action, it becomes necessary to 'level the playing field' and give the body a rest. Eliciting the relaxation response is a way to achieve that. Essentially, this consists of doing two things for a period of 10 to 20 minutes: one, focusing on a word or phrase you choose and, two, disregarding other thoughts when they come into your mind. Sounds simple? It is, but it is also difficult to do.

Here are the steps:

1 Pick a word or phrase. Choose a focus word like love, one, peace, happy, calm or, if you prefer, a religious word or short phrase that is meaningful to you.
2 Sit someplace where it is quiet and you can be comfortable but alert. Close your eyes (or leave them open if you prefer), breathe naturally and begin repeating your word or phrase.
3 If thoughts come to you, push them away and focus on your word or phrase. Continue for 10 to 20 minutes.
4 Repeat these steps a second time during the day.

Try this for two weeks as mentioned above.

Relaxation response practice
The most effective way to learn the relaxation response is to practise it. Before you begin, answer the following questions.

What is the word or phrase that you will use?

_____

Where will you do the relaxation response?

_____

When will you do the relaxation response?

_____

_____

Note in your diary or planner the times you have indicated and complete the two-week trial. When this period is over, set aside time to review the results. Writing down your brief answers to the following questions may be helpful.

What results have you observed regarding your level of relaxation?

_____

_____

Did you feel calmer immediately after each practice?

_____

_____

If so, how long did the feeling of calmness last?

_____

_____

Did you notice any change in your breathing rate?

_____

_____

Will you continue practising the relaxation response?

_____

_____

If so, note any changes that you will make in your routine.

_____

_____

If you believe that you have the basics after two weeks of remaining focused on one word or phrase and keeping out other thoughts under these conditions, the conditions may be varied. For example, consider walking or jogging, if you are physically fit enough to do that, for 20 minutes and saying 'left' when your left foot hits the ground and 'right' when your right foot hits the ground. As work or home problems enter your mind during the walk or run, push them away and continue by saying 'left/right', 'left/right', 'left/right'. Pretty soon you'll be where you want to go and will also have elicited the relaxation response.

Progressive relaxation

This is another relaxation technique that is appropriate for Bert. The technique of progressive relaxation was developed by Edmond Jacobsen, a Chicago physician, and was first published in 1929 in a book entitled *Progressive Relaxation* (Bourne, 1997). The book described 200 different muscle-relaxation exercises that took months to learn and complete. Over the years this programme has been modified by many people to result in a manageable procedure that still produces many of the same benefits.

Jacobsen's discovery was that tensing and then releasing could produce relaxation of the muscle being focused on. The process became one of awareness of the tension, then tensing the specific muscle, then releasing the muscle with the goal of muscle relaxation. The tensing and releasing went on systematically throughout the entire body and patients would experience progressive relaxation as they worked through all the muscle groups.

For a shortened version of the programme, practise progressive relaxation in a quiet place. It could be in a comfortable armchair, sofa, or on the floor. Wear loose clothing and take off your shoes. It is best to have an empty stomach.

The basic procedure is to tense a specific muscle group hard (but not straining) for 10 seconds and then suddenly let go of the tension for 15 seconds. During the relaxation phase tell yourself to let the tension go. Here is a sequence that you can follow for tensing and releasing the muscles in the body.

Remember that the key is 10 seconds of tension and then a sudden letting go and relaxing for 15 seconds. Again, plan to use this technique for a two-week trial period.

1 Clench your fists and then relax.
2 Tense your biceps by flexing a muscle in your arm and then relax.
3 Tense your triceps by holding your arms out straight and then relax.
4 Tense your forehead by raising your eyebrows as far as possible and then relax.
5 Clench your eyelids and then relax.
6 Open your jaw wide and then relax.
7 Carefully tense your neck muscles by putting your head back and then relax.
8 Raise both your shoulders up and then relax.
9 Put your shoulder blades together and then relax.
10 Tense your chest by holding in a breath and then relax.
11 Tense your stomach by holding it in and then relax.
12 Carefully arch your lower back up and then relax.
13 Tighten your buttocks and then relax.
14 Tense your thighs to your knees and then relax.
15 Tense your calves by pulling your toes towards you and then relax.
16 Curl your toes down and then relax.

Repeat the technique with any muscle groups that still feel tense. Breathe in and out easily and slowly several times. The entire procedure should take no more than 20 minutes.

Some people would rather follow the progressive relaxation procedures by audio tape. You could make your own tape or purchase a progressive relaxation tape. If you make your own tape, make sure that you leave enough space on the tape for tensing your muscle group (10 seconds) and relaxing it (15 to 20 seconds).

Practice is the key. The first few days might not bring as much relaxation as you want, but staying with it will train the muscles to relax. Learn to observe your body and be aware when you initially begin progressive relaxation not to over-tense any part of your body, particularly the neck or back, and watch out for cramping of the toes and feet.

At the end of the two-week trial period, answering the following progressive relaxation evaluation questions is useful.

What differences in muscle feelings have you observed?

_____

_____

Did relaxation occur sooner towards the end of the trial period?

_____

_____

_____

Did you observe any changes in your emotions or feelings during the trial period?

_____

_____

_____

Do you intend to continue the progressive relaxation procedure?

_____

_____

_____

Note changes in your routine that you want to make and any other comments to yourself.

_____

_____

As with many of the techniques discussed, the most effective way to learn this procedure is to try it. Decide on a time and place to do it daily. Then follow the instructions and at the end of your trial period complete the progressive relaxation evaluation. Progressive relaxation ideas are discussed in greater detail in several books and on tapes.

Clarifying 'shoulds'

Do you live by 'shoulds'? Bert appears to, as do most people. I should do this. You should do that. The dog should do this and the government or some undefined 'they' should do this or that. When our days become filled with 'shoulds' over which we have little control, our health may suffer. People may feel guilty if they don't live up to their self-imposed 'shoulds'. They may feel angry when other people do not adhere to the 'shoulds' expected of them.

Life gets very complicated if you live by too many 'shoulds'. This may be the case with Bert. If this is true, he might benefit from listing the 'shoulds' of which he is aware that rule his life. Here is one way to make such an inventory. If Bert agreed, he could use the technique specified in the following exercise.

'Shoulds' inventory

Make a list of some of the 'shoulds' that you live by at work, for example: 'I should be at my desk till late most evenings'; 'People should always manage their time precisely'.

Make a list of some of the 'shoulds' that you live by at home.

Make a list of the 'shoulds' that you hold about your boss, for example: 'She should understand that some of her demands on us are unrealistic'.

_____

_____

Make a list of the 'shoulds' that you hold about your family.

_____

_____

Review your lists and think through the difficulties that you are likely to experience as a result of the pressures caused by living with these 'shoulds'.

_____

_____

Look at your lists again and see if there are one or two 'shoulds' that you could drop. Note these below.

_____

_____

What would it feel like to live without that 'should' if you just stopped believing it? Better? Worse? Why?

_____

_____

Tomorrow, see if you can find one 'should' in your life and drop it for one day. Note the results of your experiment.

_____

_____

If Bert decided to work on his 'shoulds', it would be most useful to select one that he believes is important and occurs several times most days. At the end of his 'experimental day', it would be helpful for him to review his experience and briefly note the results of ignoring the 'should' message.

As noted earlier regarding other techniques, the best way to understand the impact and dynamics of the 'shoulds inventory' is to do it yourself. Simply follow the instructions and complete the review.

Resolution

Bert is more or less compulsive, as you noted, and somewhat a pleaser of others as well, so as you might expect he tried all three exercises. He thought that all were useful. He was so impressed by the relaxation exercises that he incorporated them in a walking and jogging routine that became a way to begin his day (you might say a ritual, but that wouldn't be kind). As for the 'shoulds inventory', he mentioned it to his wife one day. She indicated an interest in it and completed it herself. They shared their results, which actually led to a humorous exchange, and found that identifying and discussing each of their 'shoulds' was an effective and easy way to avoid or manage disagreements.

Marie: the case of the mysterious headache

Complaints

Marie is reservations supervisor for DoItMyWay (DIMY) travel agency. She has held a variety of jobs since leaving school and was hired four years ago by DIMY as a general travel agent. She moved from there to specialize first in business clients and then in international accounts. She began her supervisory position about a year ago. The job involves managing a staff of 40 agents, two assistant managers, plus several support people. DIMY's business volume varies with the seasons, so Marie constantly needs to recruit and lay off temporary personnel to accommodate business flow.

DIMY is a multi-office organization. Marie is responsible only for the regional office. She reports to two directors. One is in charge of finance and insists on careful monitoring of cash flow – namely, money received from clients and fees paid to service providers. He may request reports as often as once a day. The second director is in charge of personnel. He often needs to shift agents from one office to another to match individual office loads. Even when agents are temporarily refocused on another office, they remain physically in Marie's office, but under the supervision of the second office. Nearly all of the reservations and other transactions are done by computer, so this arrangement is technologically feasible but is often confusing regarding supervisory lines of communication. These are good times for DIMY and its business has increased 225 per cent in the year that Marie has been in her management position.

About five months ago, Marie began experiencing mild headaches (you saw it coming, didn't you?) These became increasingly debilitating and more recently have been accompanied by periods of fatigue. Marie began to treat herself with painkillers but the headaches continued. She tried to remedy her fatigue by drinking more coffee. This seemed to boost her energy level sometimes, but not consistently.

Two months ago, Marie consulted her doctor about her headaches and fatigue. The doctor ordered a series of tests appropriate for such complaints and performed a thorough physical examination and diagnostic interview, concluding from these that no abnormalities had been identified. During the interview, Marie acknowledged that for the previous three months the financial director had been making what Marie believed to be unreasonable and needless requests for special, time-consuming reports. These added significantly to her already demanding workload. In addition, she noted that the personnel manager had been pressuring her to go out to dinner or to the theatre. Even though she questioned the appropriateness of such a relationship and made excuses on several occasions, she finally accepted one invitation, hoping that it would provide an opportunity to clarify her position about socializing with superiors. It didn't. The director either did not or chose not to understand her and has become increasingly aggressive about further dating.

The doctor suggested that if the health complaints continued, she wanted Marie to return for further evaluation. She offered to prescribe mild medication for pain and gave Marie a suggested diet aimed at increasing her energy level. Then she commented that Marie herself might be the key to developing better diagnostic information. It would be helpful if Marie could establish a connection between the occurrence of physical symptoms and stressful events that she thought could be related. Marie decide to postpone the prescriptions and see what she might do about gathering the requested information.

It can be added that Marie had excellent performance reviews from a series of managers and the two current directors. Reviews take place every six months and so another was nearly due. Marie believes that there are attractive career opportunities with DIMY. She has also identified interesting positions that she is attracted to outside the company. Success in pursuing both options would depend on favourable evaluations from her two directors.

Discussion

Marie's doctor was right on target with her suggestion that Marie is the best source of clarification on her busy life and the related health complaints. Assuming that you agree, what suggestions can you make regarding how Marie might go about gaining a better understanding of the relationships between her work and her health complaints?

Sorting out Marie's life

What information do you think Marie should try to obtain?

_____

_____

What means might she use to record her observations?

_____

_____

Can you suggest a way to simplify the many kinds of information that seem to be involved?

One way to look at Marie's task is as trying to establish relationships between a range of factors. Two kinds seem important in her case. First, is there a relationship between events and behaviours at work on the one hand, and the physical complaints of headaches and unusual fatigue on the other? Second, would there be any differences in the frequency of headaches and periods of fatigue if specific conditions at work were eliminated? A mystery if we ever saw one, and that is the basis of the strategy developed below. Solving personal problems can often be done effectively by using procedures that work in other settings. For example, it is not unusual to find that the systems whizzkid does not think of applying his professional knowledge and skills to child-management problems in his own home. And how many professional problem solvers such as lawyers and physicians fall apart when faced with a leaky tap or surly clerk? Most personal problems require information as part of the solution, just as do professional problems.

In addition to clarifying the event–symptom relationships, Marie might also find benefit in using the two relaxation techniques illustrated with Bert in Case 1 of this chapter. There is evidence that progressive relaxation, the relaxation response and other techniques that relax muscles and reduce tension can remedy headaches and inexplicable related complaints. Relieving pressure on sensitive sinus and other membranes is one explanation; relaxing tense muscles is one means of reducing such pressure.

Strategy
A popular US television show in the 1950s and 1960s was *Dragnet*. The main actor was Jack Webb, who played Lt Joe

Friday on the Los Angeles police force. After a crime had been committed and Lt Friday and his Sergeant arrived at the scene of the crime, Friday often found victims or bystanders who were emotional and were blurting out all kinds of opinions and reactions. His early line at the scene of any crime was 'Just the facts, Ma'am'. Lt Friday knew that you had to start with a database built on knowing what actually happened before trying to speculate about who committed the crime. He would spend a great deal of time getting the facts before he would try to solve the crime. The same premise is often true for any individual trying to understand the relationship between a physical symptom they have and its possible cause(s).

It is best not to rely on memory to get at the facts. Memory has a convenient way of forgetting that you ate three pieces of chocolate yesterday or that you 'forgot' to walk the two miles you pledged to do every day. A daily food record would keep you honest on the first indulgence(s) and a physical activity record would note the lapse in making laps!

Given the busy to hectic work life that Marie describes, it would not be surprising if she were unaware of the variation in psychological pressure during a working day. Of course, one can always lie when keeping records, but what is the point of doing that if you are working to help yourself? In Marie's case, there is no apparent reason to think that she wants to hide unpleasant facts. And so it is appropriate to suggest that she follow the observation and recording procedure.

Record-keeping methods
The first step in record keeping is to have a method, format or form to assist you in 'getting the facts'. The clearer and simpler it is, the better. For example, if you suspect a link between headaches and caffeine, you might note in your personal planning diary at the end of a working day the number of caffeinated drinks you have had in the day and circle that number. Then you might note in another circle by the drink circle whether you had no headache, a mild, medium or severe headache during the day and how long it lasted. Keep the record of caffeine and headaches for two weeks to see if there is an apparent relationship.

Review your record at the end of two weeks. If you see, for example, that on two days you drank eight drinks of tea, coffee or cola and those were the days you had medium to severe headaches, you might decide to change specific behaviours and continue keeping records. You might suspect caffeine as the culprit and decide to consume less of it for the next two-week recording period and see if your facts support your suspicions. Are there fewer headaches? If so, you might decide to remove caffeine-based products from your shopping list for a while.

Not all record keeping can be as simple as two circles in your personal planning diary. Headaches, as a physical symptom, can be caused or exacerbated by weather conditions, the number of hours slept, drugs, alcohol, prescribed medications, food (too much, too little or allergies), stressful situations, bad moods or feelings, too much or too little exercise, high altitude, changes in body chemistry, environmental pollutants and probably a few more things. What's a record keeper to do?

Kinds of information to record

Two approaches are worth considering. The first is to keep records on everything listed above on a daily basis and to see if you note a connection between one or more items on the list and your headaches. This is a comprehensive approach, but doing it might drive you mad and cause more headaches! But if the headaches are severe enough, you might be highly motivated. The other approach is to take your top lists of suspects (just like Lt Friday) and keep records on them systematically. For example, if you suspect that weather conditions, such as changes in barometric pressure, are causing your headaches, you could track the pressure for two weeks and, like the caffeine example, also note your headache severity every day by the barometric pressure reading in your records.

Both the comprehensive record-keeping approach and the break-it-down approach have advantages and disadvantages. The comprehensive approach ensures that you get a big picture of the issue, but requires considerable daily attention. The break-it-down approach is manageable in terms of record keeping but it could take longer to solve or understand the issue of your headaches and what causes them. In either case, the purpose is to locate sufficient information regarding one or more sources of your complaints to suggest removing it or them from your daily routine. As noted, one approach is a shotgun style, the other more specifically targeted. Anyone's approach will be influenced by hunches they may have about the offending condition and their own particular style of problem solving.

Choosing a record-keeping method

Do you have a physical complaint that could be understood better by tracking it with record keeping? If so, what is it?

_____

_____

Would you prefer a comprehensive or a break-it-down approach to record keeping? Note the reasons for your choice.

_____

_____

What behaviours or conditions (such as diet, drinking, exercise, weather) do you feel might influence the physical symptom being observed?

_____

_____

Where might you record these behaviours or conditions?

_____

_____

Try now to design a simple form that would allow you to record the behaviours and conditions.

Resolution

Marie decided to focus on two kinds of event. The first kind were problems involving her staff members while they were assigned to do work for other offices. She has sufficient self-awareness to know that when she is responsible for something, she wants the control that goes with it. Sometimes it seemed to her that these temporary assignments were like loose canons on the deck, as far as supervision and control are concerned. The second kind of event she decided to record were interactions with her two directors. It was clear to her that some of the requests of both were needless and intrusive in her work routine and she knew that the personal attention from the one director was offensive and probably stress producing.

Marie designed a daily recording sheet that would unobtrusively fit into her diary. The left-hand column of the sheet listed times of day in 20-minute intervals. She divided the remainder of the sheet into three columns and headed them Reassigns, Director 1 and Director 2 respectively. Each time she had an unpleasant or stressful encounter with any of the people involved, she made a tally mark in the appropriate box or cell of her chart. If the encounter seemed particularly stressful, she circled the tally mark and made a brief note on the reverse side of the sheet. She tried not to think excessively about the recorded events at the time they occurred, focusing instead on the tasks before her.

At the end of each day, Marie reviewed her log, looking for high and low stress points and for clues that would help recall the specifics of the more serious encounters. At the end of two weeks, she had accumulated information that was useful in three ways. First, a pattern of significant stress-inducing events was beginning to emerge. For example, it was not all reassigned staff who seemed to behave in ways that were stressful to her, but only a few. Second, the information suggested that the requests for additional information from directors were related to headaches. Third, the offensive contacts with the socially demanding director were clearly upsetting, if not the cause of headaches. Marie judged the recording procedure worth the effort. The information would be useful to herself and her doctor and it increased Marie's awareness of the dynamics of the work situation from a stress perspective. As an unexpected bonus, the increased awareness led to her developing several coping techniques for managing the two negative kinds of situation. She decided to continue the recording procedure and add other suspected stress-producing events to it.

You are probably way ahead of the curve and are dying to suggest that Marie might read up on the 'shoulds inventory', and you are undoubtedly correct. That is a wise suggestion. In most stress situations there have to be two surfaces to create sparks and Marie's 'shoulds' are certainly one of them. Reviewing her 'shoulds' regarding the work place will provide a basis for determining if any are unreasonable.

Synthesizing

Bert and Marie's cases illustrate a great many concepts, ideas and procedures, yet this chapter presents the material via just one media – the printed page. Another way to conceptualize and synthesize the material is through simulation. Imagine that, as a manager, you are asked to develop a panel discussion with the aim of presenting selected points in the chapter to a group of new managers. In addition to yourself, the other two panel members are Bert and Marie. Consider how you might organize the discussion and how you could most effectively use their different experiences and personalities. Of course, you will have to make some assumptions about personalities, but the case studies themselves are a good basis. Remember to consider how you could use your own personality and experiences.

Begin by articulating the objectives of the presentation – what are the main facts, ideas and behaviours that you want your audience to understand?

You may want the panel to pay some attention to the specific problems and concerns presented by Bert and Marie, but try also to develop a broader focus on the underlying problem-solving ideas and procedures. Another suggestion is to give some thought to the questions that you would want members of the audience to raise. Whether the questions you want are likely to be raised depends, in no small measure, on the success of the panel's presentation.

Even if you don't have a real opportunity to lead such a discussion, merely simulating how you might undertake the assignment is an effective, and hopefully interesting, way of synthesizing and personalizing the material in this chapter.

4

IT'S ALL IN YOUR HEAD

Introduction

The cases in this chapter have in common the issue of personal confusion. People are confused about their professional and personal relationships and about why their current behaviours are not getting the job done. Even though feelings and behaviours are involved, their basic concern is cognitive in nature – that is, confused thinking.

The kinds of techniques and exercises described have several intended outcomes, but a main one is confusion reduction. Call it increased awareness or understanding, but their purpose is to assist people in clarifying their thoughts and understanding. The task is often to help such a person observe behaviour and conditions that may be obvious to many colleagues and friends and to understand how the behaviour affects their relationships with others. You might say that one intention of the techniques is to 'help us see ourselves as others see us'. That 'view from across the room' is not necessarily more valid than our own self-concept, but it is well worth knowing. Understanding a discrepancy between the way we see ourselves and the way others see us can be the basis for adjusting our thinking and behaviour so that we gain increased life satisfaction. The approaches here go beyond increasing awareness in that they include practical techniques for developing and creating desired behaviours.

When someone says 'It is all in your head', the implied message is that your concern is not real, perhaps even that you are imagining it or, at best, that it is simply a matter of your seeing circumstances differently from others. However, we know that those things in our heads, thoughts to be more precise, are real. The fact that no one else can see our thoughts is beside the point. We have them and they exist. So if you say 'I think I act

in a kind and considerate manner towards my staff, so I am confused about why people seem cool towards me', you are express-
ing the way you perceive the world. Your thoughts may be
inaccurate, incomplete, misguided, biased or based on faulty
information but, nevertheless, the thoughts are real for you and,
if they are upsetting or disturbing, it is important that you deal
with them. In other words, an important step is to reduce any
confusion. As you read the cases in this chapter, try to relate
them to other situations in which confusion reduction seems an
appropriate means of resolving problems.

Esther: the case of the deceptive decision

Complaints

Esther's rise in accounting had been impressive, even phenome-
nal. Her start in higher education had been a preparation for
medical school. Her academic marks placed her at the top of her
class, but the more she thought about life as a doctor, the more
she began to have doubts about its attractiveness to her. Largely
because of this, one of her tutors suggested that she do a summer
internship at a hospital. It certainly wouldn't be advanced brain
surgery, but it could give her some insight into the professional
world of medicine and also an opportunity to observe and relate
to practising doctors on other than a doctor/patient basis.

The fact that she did not enjoy the internship was what
accounted for its success. She discovered that she disliked con-
tinually caring for sick people, resented the very idea of the
medical hierarchy and thought that the restricted life doctors
seemed to live was not appealing. At the end of the summer she
was so discouraged that she chose not to return to college until
she was clearer about her career interests.

She obtained a job in the office of a large department store.
Her attractive, outgoing personality and the fact that she was
more than computer literate made her an attractive candidate
and soon she was a junior interviewer in the personnel department. As winter approached, she was looking for an interesting
activity and, almost by chance, enrolled in an evening accounting course. She immediately discovered that she had what

appeared to be a strong aptitude for whatever was involved in accounting. 'I can do this,' she said to herself.

The following autumn she enrolled in a full-time course in accountancy. Again, she finished with very high marks and was hired by a leading firm as a qualifying accountant. She continued her studies and within several years had passed all the required exams to become a fully qualified chartered accountant. During this period she had also married and had the first of her two children. Her husband, Richard, was a computer consultant who enjoyed working on his own. As they settled into this phase of their professional lives, it became clear to them both that they would probably have a more enjoyable and successful family life if they viewed themselves as a one-career family. Esther's potential for income and security seemed significantly greater than her husband's and they agreed that her career interests should be primary.

This arrangement worked well. In three years a second child was born, and it seemed natural that Richard would assume the role of primary carer. He continued his consultancy business and was able to schedule it around family responsibilities. During this same period, Esther was assigned more and more managerial tasks until at least half of her time was devoted to client relations, rather than to technical accounting work.

It was not a surprise to others in the company when she was offered a partnership. While honoured by the offer, she also saw clearly how the additional responsibility of a being a partner would influence her family relationships. However, with Richard's support, she accepted the partnership. The company had now achieved international status and was growing rapidly. This meant greater demands from larger clients, which in turn competed with personal and family time. Within a year she was working what were essentially 10- to 11-hour days, six days a week.

An unanticipated complication arose when she became ill and was away from work for five weeks recovering from surgery. It was during her recuperation that she found herself questioning the wisdom of 15 more years as a partner. There were, certainly, many professional options that could be developed, but it would be difficult to find one as lucrative as her present situation. In addition, the long-term benefits of being a partner were extremely attractive.

Shortly after returning to full-time work, Esther found that it was not unusual for her to be preoccupied with more career doubts and questions. This kind of thinking would clearly affect her work if it continued. In addition, it reinforced the value of coming to grips with her career concerns and making whatever decisions might be necessary. Richard agreed and was supportive, even to the point of saying that he would be willing to adjust his family role if that seemed to their advantage.

Having paved the way for making a career decision, Esther acknowledged that actually doing so was not as simple as it might appear. An added complication was a clause in her partnership contract that essentially placed a limit on the time during which she could resign without forgoing significant assets that she had accumulated during her tenure with the firm. To be more specific, she could resign from the partnership within the next six months and still keep company shares and endowments that she had accumulated. Once this deadline had passed, a resignation would involve selling her assets back to the partnership. Because their predicted future value was sizeable, the forfeiture would make a significant difference to her financial security. Given all of this, Esther was under considerable pressure to make a decision within the next six months.

Discussion

As with most people facing mid-career choices, Esther discovered that the many options, values and contingencies were confusing. Her decision-making case is complex and is included here because it lends itself to illustrating the many factors and components that such decisions entail. Some factors or variables are more easily defined and understood than others, especially those for which cash values can be assigned: £20 000 is £20 000 whether it is in the form of cash or a Bank of England guaranteed note. If, however, it comprises shares of stock, then the value over time is not as clear, and if it is in the form of a paid-up life disability insurance policy, then the uncertainty is even greater.

What this example calls to attention is the factor of risks in decision making. When one makes an investment with the intention and hope of making money, the risk is an obvious factor. One reason that someone is willing to pay you interest for investing your money is that you are willing to risk losing all or part of it. Your knowing the price or cost of the risk makes it easier to decide whether or not to take it.

The same concept of risk is not as easily understood when applied to variables that have no established and measurable value. In Esther's case, several intangibles are at risk in her career decision. One of these is her personal discretion over time. She could decide to leave the partnership for another option only to find that option more restrictive than her present job. Another intangible is weighing the value of more time with her family against the satisfaction that comes from being a partner.

As we will see in the following section, some risk factors and values do not become clear until people actually initiate a decision process. For example, have you ever made a holiday decision only to discover that expectations based on the brochure are not all valid? Could you have known that before spending your money and arriving at the destination? Not for certain, but perhaps you could have found someone who had decided to go to the resort and asked for their opinion. The same concept of risk is involved in many decisions. Sometimes you have to go there to find what it is really like. If that is true, then are you willing to make the decision without knowing for certain that it will meet your expectations?

The decision-making procedure that is demonstrated in the following strategy section allows you to consider non-monetary risk factors and thus is especially suited to career decisions. It is also adaptable for use in other decision-making and problem-solving situations that, left unresolved, can be stress producing.

Strategy

There is no perfect decision-making style. Some people like to be intuitive and do something when it 'feels' right; others like to turn the decision over to an authority to make for them; being impulsive or fatalistic is the style of others. Another style of decision making can be called rational and this is the decision-making style that Esther used.

Before continuing with the adventures of Esther, it would be helpful to review the basic nine steps of the decision-making strategy. You may find it useful to refer to the decision worksheet (Table 4.1) as you read the description. Don't try to use the strategy until you have read the nine steps and the following description of how Esther used it. In other words, it is best to get an overall understanding of how the strategy works before applying it to yourself.

Step 1   State your goal

Translate complaints, areas of unhappiness or concerns about your job and career into goals. Describe how you would like things to be. Think through this step carefully. Make lots of notes. Walk around the block and think about it more. After you have pondered, complained, written and thought, record a clear and specific statement of your goal in space 1. If you have identified several goals or problems, select the one that you want to work on first. Try to state your goals in positive terms. For example, 'Get away from this awful situation' is not nearly as useful a goal as 'Develop alternative job opportunities that have potential for learning and advancement'. The first is a careless cop-out, while the second reflects a positive self-directive problem-solving approach.

Step 2   Identify alternatives or options

List in space 2 every possible alternative means of achieving the goal that you can think of. If you can't think of any, ask others for suggestions and seek out information whenever possible. Regardless of how impractical or foolish some alternatives may appear, list them anyway. At this stage don't be blocked by thoughts of 'That wouldn't work!'.

Don't confuse goals and alternatives. Goals describe an end point, or at least a place or situation that we desire to achieve. In the previous example, the preferred goal statement was: 'Develop alternative job opportunities that have potential for

Table 4.1 Decision worksheet

1 GOAL				6 UNDERLYING VALUE	
2 ALTERNATIVES	3 RESOURCES NEEDED SKILLS MONEY PERSONAL OTHER	4 RISKS AND UNDESIRABLE ASPECTS	5 RISK RATING		7 VALUE RANKING

learning and advancement'. Examples of alternative ways of achieving that goal might be visiting a job centre, obtaining information from various employers, taking college courses, conferring with people already working in a field that interests you, developing a CV and reading career information. The alternatives are ways to get from here to there.

Step 3 Sharpen up the alternatives
Now review your list of alternatives. Remove those that appear redundant and eliminate any that are so much in conflict with your values that you can't accept them.

Step 4 Predict the resources
In space 3 note the resources that you would need to implement each alternative. Don't overlook personal resources such as persistence, courage, support from others and self-confidence. This step often necessitates obtaining information regarding some or all of the alternatives.

Step 5 Be realistic
Eliminate alternatives for which resources are clearly unavailable or which you believe would be too difficult to acquire. In other words, don't begin by shooting yourself in the foot. It may help to consider each alternative in terms of what you have learned about yourself from your past behaviour. Even though an alternative may seem attractive, it may not be wise to pursue it if it requires you to perform beyond your present skills and self-confidence.

Step 6 Identify risks
In space 4 note the risks and undesirable aspects that each remaining alternative entails. Risks refer to what you might lose by pursuing the alternative. Include factors such as self-esteem and relationships as well as material items. Undesirable aspects include such considerations as being inconvenienced or having to deal with unpleasant circumstances. Be honest, but don't jump the gun either. Note the risks and give yourself time to think about them. What seems risky today may not be so threatening after a few days' thought and having gained more information.

Step 7 *Evaluate risks*
Now use space 5 to rate each remaining alternative according to
your willingness to accept the risk or experience the undesirable
aspects involved. Use the following scale: 1 = acceptable; 2 =
mostly acceptable, some reservations; 3 = mostly unacceptable,
very uncomfortable with it; 4 = totally unacceptable. Rule out
all alternatives rated 4.

Step 8 *Select*
If you want to make a decision with minimum risk, select the
alternative that has the most acceptable risk level and for which
resources can be obtained. If two or more alternatives have similar ratings, go to step 9.

Step 9 *Introduce your values*
So far, we have been playing by the numbers and trying to avoid
high-risk alternatives. Now is the time to consider a higher-risk
approach. If low risk is not your most important consideration,
then you can perform an alternatives preference ranking. The
ranking is a form of 'What if?' exercise, so you can always return
to a lower-risk alternative.

To do a alternative preference ranking, first record the criteria for your ranking in space 6. Criteria vary greatly among people. They include 'positive impact on others', 'feels good',
'enhances my reputation and makes me look good', 'gets me out
of a rut and lets me try a new way of thinking about my life'. The
criteria do not have to be logical, nor do you have to disclose
them to others, but they should be thought out and clear to you.
In other words, you are giving yourself the mental luxury of not
being practical and seeing how it would be to make a decision
based on your very private desires.

Then use space 7 to rank the alternatives according to your
preference and without regard to the level of risk involved. Your
first preference will be ranked Number 1, your second Number
2 and so on. Choose the alternative that has the highest preference ranking and an acceptable risk rating.

How does that feel? Try using other criteria and see how the
rankings turn out. Take some time to play 'What if?'. This

process can clarify your goals and alternatives considerably. Often, this activity of trying various approaches clarifies the results of other steps in the process, helps you create more innovative alternatives and sharpens your self-awareness.

Resolution

Esther decided to try using the decision worksheet and follow the decision-making steps. You will recall that step 1 was stating one's goal. This may seem a bit tricky at first because decisions are often described as 'yes or no' situations. Should you do something or not do it? Should you take job A, B or C? The difficulty with that view is that you are likely to accept the limited alternatives as your only options. For example, when offered a new job, many people would frame the decision as whether or not to accept it. However, if you think about it, once you are willing to contemplate taking a particular new position, then why not expand your options and at least consider finding additional alternatives to your present job? The fact that you are willing to change jobs might tell you that it is to your advantage to look beyond the particular opportunity that has presented itself. Given this kind of thinking, Esther stated her goal as: 'To develop at least three career opportunities that meet a mix of my professional and family criteria or values.'

In step 2 she was able to develop many potential alternatives. She did so without trying to evaluate each as it was identified. The task was to get as many down on paper as she could. Mixing evaluation with broadening alternatives is often counter-productive because of the subtle limitations that you might impose on your decision-making process. For example, Esther noted 'Pursue overseas opportunities both inside and outside my company' as one alternative. Having written it, her immediate reaction was: 'Oh, that will never work. What would we do about the children's school?' Educational arrangements, of course, are a related but separate decision. Why let it dampen her creativity at this point in the process? So Esther left overseas alternatives on her list.

Step 4 is concerned with predicting the resources you will need for the several alternatives remaining after the sharpeningup operation in step 3. Several of Esther's alternatives entailed significant family transitions. An example was her moving to a new location while the family remained at home so that the children could complete the school year and her husband could meet certain business commitments. Some families would find such a transition difficult to manage. In order to assess family as a resource, she and Richard had several discussions with the children in which they examined possible situations and how they would deal with them.

As a result of these discussions, Esther and Richard decided that it would be unwise for her to use an alternative that would cause her to be absent from the family for long periods. As a result, alternatives that did not allow for her to be home for at least three days a week were eliminated. Due to the children's school schedules, this essentially limited alternatives to those that included the option of a three-day weekend. Nothing is set in stone, so the saying goes, and even this restriction could be reviewed later.

Esther identified risks for the alternatives on her list. These could be grouped under two headings: financial and family. As noted, the financial risks centred around the six-month window of opportunity that had opened. Another financial risk was the uncertainty associated with each alternative. She decided that the amount of risk could be clarified by working out the details of contracts with potential new employers. Uncertainty could not be avoided in the self-employment options. Unwilling to take that blind risk, Esther ruled out the one or two self-employment options that were still on her list. Risks associated with family concerns had been clarified in a previous step, as you saw, and she felt that these were well understood. She did remind herself, however, that commitments that a family member thought were reasonable might not turn out to have been accurately predicted once she was into the reality of a transition.

Given a list of alternatives for which the risks were all acceptable, Esther thought carefully about how her own values related to the alternatives. What was her strongest preference? One had emerged as she worked through the decision-making process. It was to find an opportunity that would allow her to use her management experience and skills, was professionally challenging and one in which she was creatively involved in developing a new organization. The two alternatives ranked first and second in terms of her preferences also had average risk factors. She now felt comfortable in further investigating both possibilities.

It is essential to highlight the importance of engaging in the decision-making process. The process itself provides a sort of synergy, in which the total result may be more than the sum of the parts. The process, in short, encourages creative problem solving.

(The decision-making material in this chapter is drawn from Loughary and Ripley, 1987.)

Norma: the case of the 'worrier'

Complaints

When Norma was a child first entering school, her mother referred to her as a 'worrier'. Her older siblings picked up on the idea and before long it stuck. Her friends, fellow students and even her teacher referred to Norma as a worrier. It never got to be an actual nickname, but if someone in a group said 'Here comes the worrier', most knew who they were talking about. As Norma grew into adulthood, she sometimes wondered which came first, the chicken or the egg, or in her case the behaviour or the label. She did think that she worried more than other people, or at least that is what they told her, and she probably worried about minor issues, or so they said. Nevertheless, the excessive worrying, as a teacher once called it, did not seem to extend to any other problems. The main disadvantage, as far as she could discern, was that she was often preoccupied with some future event or possible circumstance that might go wrong. Thus Norma more or less took her worrying for granted; for her, it was not such a big issue.

The exception to this was when she changed jobs or some other major aspect of her life and in the process encountered new circumstances and responsibilities. For example, when she finished school and was employed as an office clerk, she felt overwhelmed by everything that could possibly go wrong and would return home each day exhausted by worry. But as she became accustomed to a new job she could predict that the worry would diminish, though not disappear, because when one set of worries was over, others always emerged to take their place. The same scenario happened when she was promoted to a secretarial position and then again when she was made a PA. In each set of circumstances the pattern repeated itself and she was able to dispel job worries as she gained experience or time in the job.

Six months ago Norma applied for and was selected to fill an office supervisor position. Sure enough, she was met on the first morning of her new position by a band of worries. She was particularly worried about worrying because she had demonstrated that she was proficient in all of the many competencies involved in performing the tasks of the various office positions. All she lacked was supervisory experience. Norma worried about who might be absent, machinery malfunctions and breakdowns, unanticipated workloads, special and one-off projects and especially about planning new projects. Because planning and organizing were very important in her supervisory role, this was especially frustrating.

As Norma waited for time to increase her self-confidence and decrease her worries, she read books that she thought might help her cope with the excessive worrying. One was about time management; she didn't learn much new from it. Another was about setting goals; that didn't offer much either. She even read up on dealing with stress and that didn't seem to fit. But as time passed, the excessive worrying did not diminish as it had with past jobs. As Norma said to a close friend, 'It's really not an especially stressful position for me. In fact, I really like it and I appear to be doing very well. It is just that I can't seem to stop this infernal worrying! It is not an emotional thing at all. I just can't control my thoughts!'

'Well,' said the friend, who worked in a mental health research centre, 'Did you know that there is something called thought stopping?'

Norma smiled knowingly.

'I'm serious!' continued her friend. 'It's the truth. One of the counsellors at work uses it lots with some of our clients. He even has a little booklet on it. I'll get a copy for you, OK?'

Norma was not especially hopeful, but conceded that reading the material couldn't hurt. Anything to stop the worrying!

Me, worry?
What about you? Have you ever been concerned because you thought you were worrying too much over something? Describe the situation briefly.

_____

_____

Can you identify to some extent with Norma? Whether or not you can, think a little about the nature of worrying. That is, what does it mean to worry?

_____

_____

When should someone suspect that their worrying is a cause for concern?

_____

_____

Discussion
If by worrying we mean being concerned about important people and situations in our lives, then worrying is something that nearly everyone does (if you don't, then you may have another problem). When such concerns lead to feeling uneasy about something that could easily go wrong, then you are on the

borderline of worry. One telling factor is the probability or chance that something could go wrong. Another is the importance of the event or situation causing concern. 'Excessive' is probably the key word. Excessive concern over highly unlikely events probably qualifies as worrying. So does excessive concern over trivial or insignificant circumstances or events. The key, of course, is being able to define 'excessive'.

One suggested definition is 'concern that preoccupies your thoughts and inhibits the effectiveness of your behaviour'.

Defining excessive
Note a brief example of worry over an improbable event. Also note a brief example of worry over an insignificant event.

You can compare your examples with ours. Preoccupation over the crash of a scheduled flight of a major airline is excessive worry. While such an event is terrible, it is highly unlikely. To be preoccupied with it is to worry excessively. Preoccupation with whether or not your boss will like your new leather brief-case is excessive worry over an insignificant event. So she doesn't like it: what's the worst thing that can happen? That may be a poor example, but you do get the idea.

Instructions for two techniques for dealing with worrying about unlikely or insignificant matters are described below. These are the techniques we suggested that Norma consider.

Strategy
Thought stopping, like many of the techniques in this book, has been around for a long time. It was introduced by Bain in 1928 and then adapted by Joseph Wolpe in the 1950s for the treatment of obsessive and phobic thoughts.

Thoughts are with us all the time. When they are pleasant or neutral thoughts we hardly notice them or, in the case of pleasant thoughts, we want them to go on for as long as possible, but when they are negative thoughts it is another matter. We would rather they would go away and not bother us. Essentially, thought stopping is using your mind to control your thoughts by telling yourself to stop the unpleasant thought and replacing it with a more pleasant one. This is a good skill to learn because it is known that negative or fearful thoughts precede negative emotions. If one can learn to control the negative thoughts, the negative emotions can be diminished and thus stress can be reduced – a worthy goal.

So how do you get a handle on your negative thoughts? Start right now.

List below any recurring negative or frightening thoughts that you know you have.

_____

_____

Could you come up with some? If so, you are ready to go on to thought stopping, but to get an even better handle on your thoughts, consider keeping a negative/fearful thought diary for one day. Use the form in Table 4.2.

At the end of the day, summarize what you have learned by answering these questions.

What was your most stressful negative or fearful thought?

_____

_____

Is this a common recurring negative/fearful thought for you? Why?

_____

_____

Was the thought true? That is, by the end of the day did the negative or fearful thought come true? Most often, they don't. Describe what resulted, if anything.

Pick one negative/fearful thought to work on in the next several days and apply the following steps. Choose a thought that you would really like to extinguish. Write that thought here.

Here are the three steps for thought stopping.

Step 1 *Imagine the thought*
For practice, imagine a situation that would bring the negative or fearful thought to mind.

Step 2 *Interrupt the thought*
When you are having the thought, say out loud to yourself 'Stop!' and clear your mind of the thought. If it comes back immediately, say out loud again, 'Stop!'.

Step 3 *Substitute another thought*
In place of the negative or fearful thought, substitute a pleasant or assertive one. For example, instead of having the negative thought of 'I can't make this speech', substitute, 'I will do as well as I can when speaking'.

Table 4.2 *Negative/fearful thought diary*

TIME OF THOUGHT	SITUATION	NEGATIVE OR FEARFUL THOUGHT	EMOTIONAL REACTION	HOW STRESSFUL 1 = Low 2 = Medium 3 = High

Those are the three steps. Step 2 is the hardest to do when in public. In this case, say 'Stop' under your breath or pinch yourself or press your fingers into your palms. Some people even wear a rubber band and snap it when unwanted thoughts occur.

Practise using this technique for a week. If it is unsuccessful at first, perhaps you've chosen a very difficult thought to stop. Try it again on one of your lesser negative/fearful thoughts and build up your skill of thought stopping.

Resolution

Norma practised the thought-stopping technique for several days. At first it seemed strange, but soon it seemed natural. The technique was effective, which contributed to its seeming more natural. After a week of success with stopping negative thoughts as they occurred, Norma decided to add a more systematic component. She made a list of thoughts that consistently came to her and worked out a couple of alternative thoughts. She reasoned that having thought about these alternatives and actually rehearsed them ahead of time, she might be more successful in thought substitution. She discovered that having a reliable list of alternative thoughts was often helpful. It also made her more confident about her ability to control negative thoughts.

You might experiment with your routine should you decide to practise thought stopping. Remember that once you get the hang of it, there is no requirement that you shout 'Stop!'. You can condition yourself to use rubber band snapping or some other device with which you are comfortable.

Underlying issues

We began this chapter with the idea that confusion surrounding vocational and personal relationships can inhibit performance at work. Most people have some awareness, and experience, of this problem. When in these cases others suggest that 'It's all in your head', they may be correct, in that negative feelings are primarily the result of our own thinking and perceptions; they may also be incorrect, in that no matter what the cause, the feelings, and their impact on our health and behaviour, are real.

Two cases involving personal confusion were presented and you were offered opportunities to participate in their resolution. One way to enhance the meaningfulness of the techniques discussed is to practise applying them to your own concerns. There is no need to identify a major crisis; it may well be more beneficial to practise on smaller concerns that are more responsive to simple solutions. For example, the various components of the 'decision-making procedure' can be used independently of each other – there may be situations in which analysing the level of risk is of more concern; there may be others where developing alternatives is the key. It's possible to 'mix and match' the various ideas with specific concerns that you want to work on and resolve.

The same idea applies to 'thought stopping'. Develop your thought-stopping and thought-substitution skills by working, to begin with, on a minor but irritating thought. As with the decision-making skills, you might be able to develop and refine the general ideas so that they become specific techniques that work best for you. In other words, with the benefit of practice, develop manageable techniques that suit you and work to clear, specific objectives.

5

IT'S NOT YOU – IT'S THE WAY YOU BEHAVE

Introduction

There are times when the way we behave does not convey our intentions and so others misunderstand those intentions. There are also times when our behaviour demonstrates our thoughts and feelings clearly, but it is not to our advantage to make them known. The reactions of other people to such behaviour can be a source of personal unhappiness and frustration for us. No one likes to be misunderstood or to have temporary thoughts misconstrued as our basic attitude. The result in either case can be damaging to one's mental health.

Assertiveness is one of the most effective behavioural means of staying healthy and the first case in this chapter illustrates how aggressive, as opposed to assertive, behaviour can be a source of confusion to others. It demonstrates the connections between good mental health, assertive behaviour and effective management, and describes means for modifying our own behaviour so that it is consistent with the image we want to portray.

The second case illustrates how to deal with another behavioural condition that is referred to as boredom. Job success and a pleasant family and personal life are not a guarantee against becoming bored. Many activities, once stimulating and pleasurable, can run their course. It is not the activity that is at fault, of course, but our perception of it. Most people will probably become tired of a steady diet of any pursuit. That is one reason we change our work routines on the one hand and seek out different family activities on the other. Variety is the spice of life, or so it is said. However, we can increase our tolerance of the established routines governed by our responsibilities and commitments by initiating new leisure pursuits.

Selecting leisure activities that satisfy our interests and desires may not be as simple a process as one might think. The next case illustrates the process in a manner that allows it to be personalized to suit your interests and circumstances.

Nick: the case of the angry sleeper

Complaint

Nick is as bright as the next person. Actually, he is brighter than many and he knows it, and that seems to be part of his problem. He is a little the other side of 40, in good health and married. Two of the couple's children have almost finished school and another will do so in three or four years' time. His wife, Grace, works part-time as a dental receptionist in the neighbourhood.

Nick has been in the food business most of his working life and for the last 10 years has held management and supervisory positions with Luigi's Fine Italian Cuisine chain. Luigi's has stores in all the larger towns. Nick has been assistant manager in several smaller stores, worked in the training department for a period and had a stint as a supervisor in Luigi's frozen-food plant for another. He has held one of the two assistant manager positions in the Park Centre store for nearly three years. During that period, four assistant managers have moved on to manage smaller stores of their own. This is essentially the career path of managers at Luigi's.

It was clear to Nick when he was transferred to be an assistant manager at Park Centre that he was finally on the way to becoming a senior store manager. In fact, so certain was he about his future that he and Grace celebrated by taking a week's holiday in The Netherlands. That was almost three years ago. To say that Nick is discouraged these days would be a gross understatement. He is more confused than depressed about what he considers to be his poor fortune. Grace has observed that for about two years Nick has been irritable at home. He is less patient with the children and tries to resolve domestic difficulties by 'administrative edict'. Within the last six months, Nick has also experienced difficulty in sleeping through the night. It has reached the point where he is sleeping soundly for fewer than five hours most nights. Grace knew that sleeping was once a skill at which he surpassed and so, with this turn of events seeming to her to be the final straw, she asked him to talk about his worries. At first Nick denied that he had any, but as his behaviour persisted so did his wife.

Finally, one Friday evening when all the children were away, Grace confronted him seriously: 'You may think that whatever is bothering you is your business, but you are really beginning to get to me. So, please talk with me about whatever seems to be the problem. If it has nothing to do with me, then at least I can get on with my life. Now come on, Nick, what is the matter? It must be something at the store.'

Nick became defensive and began to raise his voice, telling his wife that it was, in fact, none of her business. Grace would have none of it, insisting that he give her some explanation. Nick sighed and said that she was correct. He was more and more confused and upset over his lack of promotion, so much so that he was having difficulty sleeping. He reminded her that his quarterly performance reviews had been good to excellent for several years. He had been pleased that they had continued in this manner since his transfer to Park Centre. 'That's all good,' he said, 'But on the other hand I keep getting passed over for promotion to manager of my own store.'

'Why not meet with your manager and ask for clarification?' Grace suggested.

'Well, up to now I thought that might seem like complaining. That could work against me in the long run.'

'But you might get something clarified,' Grace replied. 'You've got to know where you stand. Being in limbo is not fair to you or the family.'

Getting clarity

What would you advise Nick about approaching his manager for clarification? What would you suggest he might do to prepare for the meeting? If you don't think that he should approach his manager, what alternative for gaining clarification can you suggest?

Even though he was only half way between his six-month performance evaluations, Nick decided to ask Tom, his manager, for a meeting to clarify his situation. Tom had been manager of the Park Centre store for six months. Nick felt that he got along well with him, although his last performance review had been given by a previous manager. Tom readily agreed to meet with Nick.

Nick began the meeting by explaining his confusion surrounding the discrepancy between his strong performance evaluations and being passed over for promotion. Tom thought a moment and then said, 'Nick, I had planned to have a discussion regarding your career progress in the company at the next performance review. But since you have raised the issue, it would be just as well to have the discussion now. I have been concerned about your situation since I became manager and have reviewed your records and talked with several of your former supervisors. Maybe a good way to proceed is for me to summarize how I see your situation and then let you respond. OK?'

Nick agreed and, while not knowing what to expect, was glad that somebody was finally about to explain his situation.

'First of all, as you know from your evaluations, your performance in the technical aspects of your job is excellent. The numbers are all there. Your paperwork is excellent, for example, and your stock control is never in question. You are, in a word, very good technically. Please...,' Tom continued, as it was apparent that Nick wanted to respond, 'let me finish. Then you can have your say and we can discuss things further.'

Tom explained that over the last five years Luigi's had instituted a policy of requiring a degree qualification of prospective entrants to the management programme. Tom, and several others who had already been taken on before then, had been allowed to continue along the management career path. Tom indicated that while the company did not want to be unfair to older managers, it was also true that those without a degree did not fit the ideal profile and that might be a factor to consider in Nick's case. More important than that, Tom continued, was the perception of past supervisors that Nick manifested an aggressive manner or style of working with staff that was out of line with the company's desired management style.

Nick answered that yes, he did have a direct, no-nonsense style. He set high standards for himself and expected his staff to meet them as well. Further, he added, he thought that allowing inefficient and sloppy work to continue was harmful to the company in the long term.

Tom agreed with Nick's principles, but suggested that it might be his style of interacting with his staff that was the important concern.

Nick immediately disagreed and said, in a somewhat heated manner, that style had little to do with efficient performance. He just didn't understand why someone with his high technical performance wasn't getting promoted.

'All right,' said Tom. 'Let's take your response right now as an example. You have a perfect right to disagree with my comments, but you went about in it an aggressive and ineffective manner. Can you see that you might have two objectives in our conversation? One is to make a point about your disagreement with my comments, and another is to maintain a good working relationship so that our meeting will continue in a way that is to your benefit. Just to emphasize the point, if you are offensive to me, you lose your larger case. That may not be fair and, by the way, I'm just using this incident as an example, but it is how things are.'

Tom summarized by noting that because of Nick's age, lack of a degree and history of being passed over for promotion to manager, he was not in an enviable position in the company. 'When top management looks at you they see a technically competent, older assistant manager who doesn't fit the desired educational profile, and who seems to become more aggressive as time goes by. In order to remain on the management track, you will need to learn how to deal with people in a less aggressive, but effective, manner. If you can do that, there is a good chance that you can meet the goal you have set for yourself. Perhaps the company can help you. Why not think about what we have discussed? Let's talk again on Monday.'

Discussion
Nick got an earful, didn't he? He is certainly between a rock and a hard place. He has several alternatives open to him. Here are three. One, he can try to be less aggressive, whatever that means, and keep up his otherwise good work. His hope here is that hard work will save the day. Second, he could continue to perform well as an assistant manager, but immediately begin to search for opportunities outside the company. His thought here is that he has already been labelled an 'also ran' at Luigi's and so he would probably be better off using his skills in a new, less biased environment. Third, he might ask Tom for help from the company in learning to become less aggressive, even though that might put him in a vulnerable position should he not make the desired changes. His hope here is that the company wants him to succeed.

Your choice
Put yourself in Nick's position. Which alternative would you choose or which additional one is more appealing? Summarize your reasoning.

After some thought, and having discussed the situation with his wife, Nick decided to pursue the third option. After all, the company had been supportive to him for many years and he had no reason to distrust them now. He had a gut feeling that Tom was sincere and could be trusted and, whatever programme he might have in mind, it probably couldn't do any harm. Nick, to tell the truth, was still confused about this aggressiveness stuff. Weren't managers supposed to be aggressive?

Strategy

Nick decided that it was up to him to change. Tom suggested that he could start learning new behaviours by reading and working through some open-learning material on assertiveness. He also offered to discuss the material with Nick, who agreed.

Nick decided that he would read and discuss the material with three questions in mind. The first was: 'What is assertive behaviour and how does it differ from non-assertive behaviour?' The second question was: 'How does assertive behaviour differ from aggressive behaviour?' That seemed to be a key issue for Nick, because his inability to distinguish the difference was what was standing in the way of being promoted. His third question was: 'What are some assertive behaviour skills that I can practise?'

Here is a sample of the material that Nick read regarding assertive behaviour.

Before reading the material below, note what you think are key features of assertive behaviour.

When we are assertive, we tell people what we want, need or would prefer. We state our preference clearly and confidently, without belittling ourselves or others, without being threatening or putting other people down. Assertive people can initiate conversation, can compliment others and receive compliments gracefully, can cope with constructive, justified criticism and can give it too. It's a positive way of behaving that doesn't involve violating the rights of other people. Above all, assertive behaviour is appropriate behaviour. For example, it can mean that choosing not to be assertive in a particular situation, or with a particular person, is OK.

Assertive behaviour is first of all purposeful behaviour. In order to be assertive, therefore, we must have a purpose in mind. In the world of managers, there are usually several purposes involved. For example, you may want your people to be effective, responsive, imaginative, efficient, patient and friendly. This means that you must be thoughtful about both what you say and how you say it.

For example, telling an employee to 'get the job done as best you can' gets both you and the employee into trouble if you actually prefer (or insist on) it being done in a certain manner. To be assertive would be to say: 'Betty, it is important to do this job as described in our manual. Please be sure that you understand the procedures and feel free to consult with me if you have questions. I'm interested in taking time to confer so that you do the task correctly.' Of course, if you make such a statement with a sneer on your face, or in a sarcastic manner, you are not being assertive. Some might say that you are acting foolishly because your behaviour is at odds with your purpose, which in this case is to help Betty get it right.

The way we act physically tells people a lot about us. Our body language often reveals how we feel and may show uncertainty or contradict what we say. A good example is a manager who says to a staff member, 'I'm not angry about the way you handled that customer', but whose clenched teeth, rigid posture and grim expression reveal that they are really upset. Non-verbal language often contradicts the spoken word.

An example

Describe a personal example of being on the receiving end of a mixed message (that is, the speaker's words saying one thing, their body language saying another). What effect did/does that kind of experience have on you?

In addition to being purposeful, assertive behaviour is not meant to be offensive. Your staff may not like your instructions or the responsibility they entail, but being assertive should not give them legitimate cause to react in a negative manner to you. One way to keep your behaviour inoffensive while effective is to avoid being personal. Focus on the requirements of the task or job for which you both have a responsibility, rather than on your own desires.

For example, saying 'I want this report by 3 o'clock' may be a true statement of your desires, but also offers the member of staff an opportunity to think, 'So what?'. In other words, making the member of staff aware of your wants is really not the key issue. It would usually be more assertive to say something such as 'We need your report by 3 o'clock. It is an important piece of the project and our group's success depends on having your work.' Clearly, if the employee perceives that this is really not true and that you are primarily interested in what is most convenient to you, then your behaviour may work against you. You may get the report, but your approach may have contributed to developing a negative, non-trusting relationship with that member of staff and, because one of your purposes as a manager is building trust, being perceived negatively in this instance works against your purpose.

Nick and Tom met briefly to discuss the reading that Nick had been doing. 'I think I understand that assertive behaviour is both purposeful and non-offensive,' Nick offered. 'But how does it differ from non-assertive behaviour? Must managers always be preoccupied with purpose?'

'I suppose we must, to some degree,' Tom answered. 'That is one of the things we are paid to do. But preoccupied may be too strong a term. At one level, being alert to potential ways to be helpful and prevent problems may be more to the point, but simply behaving in a natural manner much of the time is OK, too. Perhaps a good distinction is that of active and inactive states. When you are in the active mode as a manager, it is usually important to be assertive. That is, it is important that people understand what you want done and your concerns regarding both results and methods. Obvious examples are when you are conducting a meeting, doing training, and evaluating. However, when you are in the inactive mode, assertiveness may be inappropriate and work against your developing good staff relations. An effective manager can be respected for their professional ability and liked for the kind of person they are.'

Nick smiled and said that he got the point.

Active versus inactive
Describe a couple of examples of your own 'active' and 'inactive' modes and how you want to be perceived by your staff when you are in these modes.

Nick continued his reading. Following is a sample of what he read regarding aggressive behaviour.

Before you continue reading, what words come to mind when you think about aggressive behaviour? Write down all the words you can think of to describe aggressive behaviour.

Do your words for describing aggressive behaviour refer to body language, tone of voice or other attempts to dominate the situation? How do you respond to aggressive behaviour? Most people are not attracted by it. One reason that people find aggressive behaviour distasteful is that the focus is exclusively on the aggressor's desires; another is that the behaviour itself is offensive.

Aggressive behaviour is the kind that expresses feelings and opinions in a way that punishes, threatens or puts the other person down. The aim of this behaviour is for the aggressor to get their own way, no matter what. When we are sarcastic, manipulative, when we spread gossip or make racist or sexist remarks, we are behaving as aggressively as when we bully or push. If we win and get what we want, it probably leaves someone else with the bad feeling that they have lost. Aside from any ethical considerations, this could auger badly for us in future transactions with that person. Another possible consequence of behaving aggressively is that we might feel guilty later.

Aggressive people often stand too close to others or they stand when others sit, they point or wag their fingers and use a loud, hectoring tone of voice. They may pat people (on the shoulder for instance) which is patronizing and reduces the other person's status.

Now let's focus on aggressive behaviour in the workplace, particularly turning the spotlight on manager behaviour and your personal experience.

Complete each of the following instructions to a member of staff using the aggressive sentence beginnings:

You must..

You should..

You ought to...

You've got to...

Why haven't you...?

In the space opposite, restate each sentence in an assertive manner.

Hopefully, it will be apparent that the assertive instructions are clear, not loaded, not domineering or putting down the other person. Assertive statements focus on the tasks to be done and the responsibility that you and the member of staff share. If this is not apparent, try to rework your assertive statements so that they are both purposeful and share responsibility.

Any manager can feel aggressive when threatened, frustrated or disappointed by an employee. Aggression can also stem from a manager's own feeling of failure or frustration – the old 'kicking the dog' reaction. When feeling aggression is manifested in aggressive behaviour, it is usually a sign that managers have lost their cool, control or perspective. The cause may be employee related, or simply that they got out of bed on the wrong side, or something has gone badly with their lives. If you find yourself resorting to aggressive behaviour, ask yourself what is wrong and why you are projecting your bad feelings on to someone else. Take time out to think of more constructive ways of handling whatever has gone wrong.

Your aggression
Describe a supervisory situation in which you 'lost it' and became aggressive or saw somebody else do so. How do you account for it? If you, or the other person, were to re-enact the situation to get a better outcome, how might you/they do that?

Nick conferred again with Tom about his reading. He said that he had a clearer understanding now of how he was perceived by his staff as aggressive, when what he intended was to be an effective manager. He acknowledged that, regardless of his intentions, he could see why his employees viewed him as aggressive in his methods. For example, Nick said that on reflection he realized that the more concerned he became about a person's error, the louder the voice he used. He realized also that the more concerned or frustrated he became, the more aggressive was the body language he displayed, including waving his arms, gesturing and hitting a table or some other surface for emphasis. He thought too that he seldom began addressing a performance issue by asking pertinent questions about the individual's thoughts and explanation before starting an aggressive description of a solution. It would be fair, he acknowledged, to describe his style of management as the 'attack' mode. Clearly, he had much to learn about becoming more assertive and less aggressive.

Then Nick turned to his final question – that concerning assertiveness skills that he could practise. Tom said that there were no perfect formulae that would work for everyone, but that he could provide a list of assertiveness skills from which Nick might choose some. Below is the list that Tom provided.

Assertiveness skills
1 Know what you want to say. You won't appear confident if you are unsure of what you want. You could appear foolish by asking for something that you eventually realize is not what you want.
2 Say it. Don't hesitate or beat about the bush, come right out! Practise before you say it and check for appropriateness.
3 Be specific. Say exactly what you want or do not want of a member of staff, so that confusion will be minimal. No long explanations or excuses are necessary.
4 Brief reasons for requests and expressions of appreciation and praise are usually effective motivators.
5 When discussing an oversight with a member of staff, do not let too much time pass, as this builds apprehension. On the

other hand, don't try to be assertive if you are angry.

6 Look the individual in the eye when you speak. Most people feel more comfortable if you look directly at them.

7 Look relaxed. You will convey anxiety if you shift from one foot to the other, wave your arms around or, conversely, are too rigid. Practise looking relaxed in a mirror – this is not as contradictory as it sounds.

8 Avoid laughing nervously. Smile if it's appropriate, but if you giggle or laugh you won't look as if you mean what you say. This may confuse a colleague.

9 Don't whine or be sarcastic. Be direct and honest. Whining and pleading can annoy, confuse or generate guilt feeling in the employee. This is also manipulative behaviour. Being sarcastic communicates hostility, as you are putting the other person down.

10 Be direct and honest in your dealings as a manager. Don't play games or try to be 'cute'.

Assertiveness practice
Note here the skills that you feel reasonably confident about and think you can use.

Let's try to practise all these dimensions of assertive behaviour by putting together the different components of the positive process. Think of a management situation in which you would like to be more assertive or less aggressive. Take your time and visualize the scenario as you complete the lines below.

1 First describe the situation. Know what you want to say. My purpose regarding...............................is:

---

2   Say it! The sentence I will use is:

_____

_____

3   Be specific. Review your sentence and revise it if it is not clear.

4   Say it as soon as possible. Choose your time and place appropriately. Choose a setting that will help you feel at ease and able to use the body language we have discussed – but don't put it off! I will speak to..............................
on................................

5   After you have done your assertive bit, think about what happened and what you learned from the exercise.

_____

Resolution

Nick spent several evenings working on the assertiveness material. As he progressed, he discussed with Grace what he was reading and thinking. When he had finished, she asked him what he thought about it and what it meant for his future at Luigi's. Nick said that he was pretty clear about the difference between assertive and aggressive behaviour and had some useful ideas about being more assertive and less aggressive. He said, too, that he could now understand how his behaviour might be interpreted as aggressive, when he simply meant to be very results oriented, forceful and on target. The line is not always clear, he thought, and he added that one's basic personality was an important factor in appearing genuinely assertive rather than aggressive.

'So, what is your next step?' Grace asked.

'It may be a long shot,' Nick said, 'but here is my plan. I'm going to schedule a follow-up appointment with Tom. I'm going to raise a couple of critical issues. One is why no one from the company has been straightforward with me about the aggressiveness business before this. Another is for us to agree on some goals and how and when I can demonstrate that I am getting the message. I really think that I deserve a manager's position, but I know that demanding it will not get it for me. I plan to be effective in an assertive manner.'

Ed: a case of the bored success

Complaint

Ed is service manager for Anglo motors, a large car dealership. He has a good background in motor mechanics and 12 years' experience, the last three as a manager. During that time Anglo has sent him to many special courses both in the UK and abroad. Ed has a pleasant, outgoing personality and a reassuring manner. He also has good people skills, which are invaluable in interacting with colleagues and customers alike and contribute to both low staff turnover and an enviable list of loyal customers. The managing director of Anglo recognizes Ed's value to the company and makes sure that his salary remains competitive. Ed is aware of this and knows that he is probably the highest-paid service manager in the area. The managing director has also assured him that as the company grows it will undoubtedly enlarge its service function to divisional status and that Ed has an excellent chance of being its first director.

Everything considered, Ed knows that he is currently in an advantageous position with excellent prospects for the future. The only limitation on his career development rests with the size and growth of the company. He has received offers from competing companies and, after evaluating them, decided that they would not advance his current or future career. His personal and family life are satisfying. Yet Ed has become increasingly aware that he would enjoy additional challenges and opportunities. His discontent is not a complaint, but he realizes that without additional intellectual stimulation of some kind he runs the risk of becoming a less interesting person, which in turn may have negative implications for his career and family life.

Ed considered signing up for some job-related courses but thought better of it. While he does well in job-oriented short courses, his earlier experience of courses taught him that he did not do well in general academic situations. 'Give me something I can get my hands on,' he would say. On the recommendation of a friend, Ed visited a counsellor at a local college of further education. The counsellor seemed to understand Ed's interests, as well as his concern about enrolling in evening study courses. He also suggested to Ed that it is possible to combine recreational and educational activities and that this might suit Ed at this time in his life. A useful next step, the counsellor suggested, would be to complete an exercise called 'selecting leisure activities'. Ed agreed to do that.

Discussion

Ed is a good example of a person who can benefit from an in-depth look at his leisure interests and find alternative ways of satisfying them. Up to this point, his job has apparently provided sufficient stimulation and personal reward. As he reaches a point where he would like to broaden his activities, exploring leisure activities is an attractive alternative to trying to wring more satisfaction out of a job than it can reasonably provide. Ed might follow the example of some of his colleagues who more or less stumbled into satisfying leisure. However, this is probably more apt to happen with regard to more recreational leisure than with the stimulating activity he is seeking.

People searching for leisure with greater substance or depth may benefit from a systematic analysis of what 'pay-offs' they want and get from leisure activities. The selecting leisure activities exercise has several features. One is that it helps people make the distinction between activities and pay-offs. A pay-off, in leisure terms, is a return or reward that you get from an activity. An activity is a particular way of achieving that pay-off. For example, let's say that you are interested in learning more about a particular topic and that you would like to have some tangible results from whatever studies you take up. What leisure activity might satisfy that interest? Building a collection of something might. Ed, for example, might think through this option and decide that it could be really interesting for him to collect antique cars. That could be one answer for him, but then he might investigate the costs of such a hobby and conclude that collecting cars is just too expensive for someone of his means.

But hold on a moment! Why does it have to be antique cars? Would collecting or studying any of the following satisfy Ed's interests: model antique cars, films or videos of antique cars, books about antique cars, movies featuring antique cars, famous people who had antique car collections, the role of particular cars in history or the role of antique cars in women's fashions? Or are antique cars necessarily the answer? Get involved in this kind of thinking yourself. Consider the pay-offs that you associate with building a collection of something. Pay-offs might include contact with similarly interested people, building up expertise and a reputation, pride in owning a collection, stimulation from researching a subject and a purpose for travelling. Do you get the point? It is the pay-off that the activity provides that gives it meaning and value. No leisure activity has meaning or value in itself. Those come from the benefits or rewards they bring to the person involved – the pay-offs.

Another useful result of working with the selecting leisure activities exercise (SLA) is that it can identify activities that can satisfy more than one interest; that is, provide several pay-offs. The best way to understand the SLA exercise is to do it for yourself. In addition, in the last part of the strategy section we follow Ed through the procedure and find out what it did for him.

Strategy

Selecting leisure activities
Leisure – the time for doing what you want to do. Many hard-working people have difficulty arranging for quality leisure time. This exercise can help you analyse your leisure desires and then select activities that meet those desires. Many people believe

that leisure should happen spontaneously without forethought and planning, but this is usually not true. Planning often helps spontaneity or at least ensures that some leisure activity takes place.

 While we all have a variety of desires or wishes at any point in our lives, we are not always fully aware of them. Below are 13 desires that people give as reasons for engaging in leisure activities. As a means of clarifying your own desires, read the list and tick those that appeal to you.

◆ **Desire to meet other people.** A leisure activity can be a way to meet other people. You have just moved to a new town, you are looking for a new relationship or you just want to enlarge your circle of friends.

◆ **Desire to earn extra money**. A look at your monthly budget necessitates the consideration of leisure activities that help bring in some income. Gardening for others, renovating and selling cars or making children's toys for sale are examples of leisure activities that can raise money.

◆ **Desire to enhance your contacts with your family**. The family that plays together stays together. Many families choose leisure activities that will assist in strengthening family relationships.

◆ **Desire to produce a product**. To some extent we all like to see results. If, in your opinion, you are not producing anything notable in your job, you might want to start weaving, renovating antiques, woodcarving, painting or something else you can point to and say, 'See, I did that.'

◆ **Desire to use intellect, talent or to feel creative**. Many people have talents for which they will never be paid but which can be pursued at leisure. The advantages in using musical, artistic, dramatic and intellectual talents on an amateur basis is that you, not your employer or manager, can choose the activity and influence the quality of what results from it.

◆ **Desire to help others**. We usually feel better about ourselves, and the world in general, when we help others. The desire to help can be satisfied by a great variety of activities.

These include volunteering to help in a hospital or community agency or working for a campaign or political party.

◆ ***Desire to compete.*** We all enjoy proving ourselves in some form of competition. Leisure activities can provide the excitement of competition without the risks that one might face in being competitive at work.

◆ ***Desire to appreciate nature.*** Nature and the rewards it can offer in peace, solitude and scientific wonder are hard to match. Responding to the call of the wild can bring a better balance to life.

◆ ***Desire to escape.*** There are times when escape from some unpleasant or boring reality is our priority. A stressful relationship, a work role or the aggravations of parenting are examples of situations in which escape to some completely different, even frivolous, activity may be the most beneficial reward from our leisure.

◆ ***Desire for aesthetic enrichment.*** As people come to appreciate art, music, drama and other cultural experiences, the desire for aesthetic enrichment grows. The satisfaction derived is a very personal experience and one highly valued by some people.

◆ ***Desire to be entertained.*** Not as lofty as the previous desires, but a really attractive aspiration for most of us. Activities such as cinema going, listening to our favourite kinds of music or spending an evening watching comedy or a show can fulfil the desire for entertainment with minimal demands on one's energy.

◆ ***Desire to be different.*** We all like to believe that we are unique. If life seems routine, you might want a dramatic or at least a different type of leisure activity. You might choose trekking in Nepal, scuba diving or pot-holing in order to be different.

◆ ***Desire for exercise.*** Passing a certain birthday or an early morning look in the mirror makes many seek exercise. A common problem is finding a form of exercise that is also satisfying. Combining this desire with another (such as appreciating nature) is one solution. Walking by the sea or through attractive countryside can provide double value.

From desires to activities

Let's turn your desires into potential leisure activities. Review the list of leisure desires that you ticked and select your three strongest desires. Write them in Table 5.1. Then under each desire, list specific leisure activities that would meet that desire.

Table 5.1 *Selecting leisure activities worksheet*

DESIRE 1	DESIRE 2	DESIRE 3
ACTIVITIES	ACTIVITIES	ACTIVITIES

After you have listed a range of leisure activities under each desire, look at your lists again. For the purposes of further investigation, circle one activity from your entire list that for whatever reason particularly appeals to you.

Now look at that activity carefully. Answer the following questions about this activity.

How much time does the activity take? Could you fit it into your current lifestyle?

How much initial investment is involved? Include ongoing costs. Does that fit with your budget?

_____

How much space, if any, does the activity require? Can you meet the minimal requirements?

_____

What aptitude, skill or training is required for the activity? If you don't possess the necessary skills, how can you acquire them?

_____

How will this activity affect other people in your life? Can you cope with others' reactions to your new interest?

_____

If you can meet these requirements, start planning now for a new leisure activity. If not, select another activity from your list and see if it is more suitable.

The pay-off for you will be a leisure activity that meets your desires and for which you can meet the necessary requirements.

Resolution

Ed surprised himself. Two of his main desires were increasing the quality of family contacts and to appreciate nature. His wife shared these desires as well. They discussed these mutual aspirations and realized that they could have family travel as an activity. At first, Ed and his wife considered travel too demanding and confining for a family leisure activity. After more thought, and having reviewed trips that they might make and discussed them with the children, they discovered many potential short day outings that could easily be made and would test the feasibility of their interests. Ed noted that he probably would not have developed this scheme without the structure and tips provided by the SLA worksheet.

Ed's leisure activities and the pay-offs they brought produced new balance and satisfaction in his life generally.

More practice

This chapter has included opportunities to record your responses to several ideas and suggestions. You can also practise and experiment with some of the ideas more actively. All you need is a mirror; a cassette recorder is also useful. The procedure is simple – put some of the suggestions into practice, out loud, in front of a mirror. You may want to record, and play back, your practice sessions.

For example, experiment with the aggressive/assertive sentence-completion exercises. Experiment with different endings to the sentences, using varying degrees of aggression. Don't feel shy – exaggerating differences will lead to a better understanding of the range of language that can be used and its effect. Try adding a smile, vary the tone and volume of your voice and the speed at which you talk. Eventually, try to begin developing a more assertive style that seems effective and with which you are comfortable. When you are more confident, practise your approaches in real situations and carefully observe others' reactions.

You can practise the 10 assertiveness skills in a similar manner. Don't be embarrassed – you may be self-conscious when you begin, but keep practising until you feel that you have more control over your expressions.

You can also extend the selecting leisure activities technique with additional practice. For example, try combining different activities. Be creative in identifying less expensive resources and less obvious alternatives. Search out existing groups that enjoy similar activities and consider joining them, whether at work or outside it. Eventually, you might want to try extending yourself to activities offering increased challenge and diversity. Also examine the possibilities presented by leisure activities that support your goals at work. While one of the attractions of leisure is that it is different from work, that doesn't mean that some of your leisure can't be work related; in other words, enrich your leisure so that it enhances satisfaction in both your work and personal life.

6

MIND/BODY EXERCISES FOR THE EXCEPTIONAL MANAGER

Introduction

This chapter explores six areas of mind/body health that are important to the overall well-being of those with significant work responsibilities. Managers are only as effective as they are healthy. Certainly, people who are unwell or unfit can also be effective managers, but that would hardly be one's preference. The health components explored here are psychological outlook, social support, diet, physical exercise, alcohol consumption and other behaviour and concerns.

Chapter 6 changes the format from case study to information generation. Each section considers one of the areas of focus and provides a means of collecting information regarding your own status or condition. Questions for discussion and further analysis are included. You might view these exercises in self-awareness only as an introduction. You can take each topic further by pursuing additional materials that help increase self-awareness. Your responses might also be clues to further development and changes that you might wish to make in these areas of your life. As always with such exercises, there are no right or wrong answers. Where feasible, we have provided comparison points but these are meant only to be descriptive, not evaluative. Being at the top of a comparison group of managers, for example, doesn't necessarily imply that you are also among the best. You may be, but that is for you to decide.

Psychological outlook

There is mounting evidence that the self-fulfilling prophecy is a significant factor in determining health. People with positive, optimistic outlooks, who feel that they are in charge of their lives and who expect to be healthy, generally remain healthy even after periods of intense stress. On the other hand, people with negative, pessimistic outlooks, who feel that they have no control over their lives and expect to become ill, generally experience more illness even if stress levels remain moderate. In a nutshell, we become what we think about – what we expect to happen tends to happen.

To assess your psychological outlook, place one of the following numbers beside each of the 15 statements in Table 6.1 overleaf.

1   = Among the most effective one-third of managers I have known.
2   = Among the middle one-third of managers I have known in terms of effectiveness.
3   = Among the least effective one-third of managers I have known.

After you have completed the table, answer the following questions.

What has occurred to you about your current situation as you worked through that checklist?

_____

_____

What changes might you make to that situation?

_____

_____

Table 6.1 *Psychological outlook questionnaire*

In relation to other managers:	
1 How would you rate your present health?	
2 In general, how satisfied are you with your lifestyle at present?	
3 If you make no changes in your present lifestyle, how likely do you think it is that you will develop a serious illness in the next five years?	
4 How often do you feel a sense of fulfilment and accomplishment at the end of the day?	
5 To what extent do you feel responsible for your present situation and level of well-being?	
6 How regularly do you strive to become more knowledgeable about yourself?	
7 How adequately are you supported by your spiritual or religious beliefs?	
8 To what extent is your life full of interesting and exciting things?	
9 How frequently do you feel optimistic and confident of your abilities?	
10 To what extent do you treat others with care, sensitivity and respect?	
11 How much value do you place on your feelings?	
12 How readily do you acknowledge and share your feelings with others?	
13 How much of the time do you feel that you are in control of your life and what happens to you?	
14 How much importance do you attach to your present work?	
15 To what extent does life present you with interesting challenges?	

How can you use the results? In general, any item marked 3 is probably worthy of further thought. What are your reasons for putting yourself among the lowest one-third of your peers? What evidence are you using? What does the low mark mean? Is it important to you? Would you like to change your psychological outlook status? How might you begin to do that?

You could ask some of the same questions about items in which you placed yourself in the top one-third of your peers. Check your validity. How do you know? On what behaviour and other evidence are you basing your placement? Would it be meaningful to use other criteria? If you are pleased with your status, what can you do to maintain it?

You probably answered most of the items from a general perspective. It might also be interesting and helpful to answer them from a more specific work perspective and also as viewed from your non-working life and then compare the three sets of answers. Often one context of our lives overshadows another with regard to psychological perspective and that may cause us to make unreasonable generalizations. For example, a person might put themselves in the bottom one-third regarding item 15 if viewed from a work perspective and overlook the high level of satisfaction they have in their personal life.

If you are interested in knowing more about your psychological outlook, try keeping a diary or some other record about the items that interest you.

Finally, for your answers to items of particular interest, ask yourself the two basic questions of self assessment:

◆ What do you mean?
◆ How do you know?

The kind of things that most of us need in order to manage our stress levels more effectively and to improve our health reserves are relatively straightforward and yet many people seem unable to carry through their intentions to improve. The new exercise programme lasts only a few weeks; the lost weight is regained; the time-management practices are gradually forgotten or declared to be unrealistic. This section is intended to explore why this happens so frequently.

Our *perceptions* determine what we actually experience in a situation, by making us receptive to some stimuli and blind to other stimuli. People who expect to find problems in a given situation are usually able to find them, while others who expect to find opportunities in the same situation will actually be successful.

The *experiences* we have determine our outlook. Beliefs and attitudes are formed as a result of the repetitive thematic messages that we get from outside authorities, as well as those we give ourselves by means of our inner self-talk. If we repeatedly learn from our parents and teachers that we are 'not ready yet', we are likely to develop a cautious, 'wait and see' outlook on life.

Our *outlook* determines what we perceive in a situation. A person who has an optimistic, positive, self-directed outlook will generally expect to find opportunities, while another who has a pessimistic, negative, disempowered outlook will generally expect to find difficulties.

And, as Figure 6.1 indicates, we have come full circle.

Figure 6.1 *Perceptions–experience–outlook*

In this way, our general outlook becomes self-fulfilling and self-reinforcing. In order to be fully successful in implementing our plans for improved stress management and health, it is likely that each of us will have to break into this cycle to create new beliefs and expectations that support the changes that we wish to make.

How would you characterize your own general outlook?

For each negative thought pattern that you identify, create a reverse image of the pattern and, for the next few days, practise holding this positive new image in your mind as if it were already true. The greater the number of times you can dwell on this desired new image, the more quickly it will become a part of your revised general outlook! Repeat this exercise for each item for which you scored 3 on the psychological outlook questionnaire. You will be amazed at how much your overall outlook can be changed when, with practice, you take control of it.

Social support

People vary a great deal regarding the number, type and intensity of preferred social relationships, all the way from 'none' to 'never enough'. While the evidence suggests that people who have a good network of social contacts are probably healthier than those who don't, there is not necessarily a causal relationship between the two items. Nevertheless, if social relationships are of any consequence to you, the following exercise and discussion material may be useful and interesting.

Managers live a lonely life, it has been observed, and the higher on the ladder the greater the loneliness. There seems to be some truth to that and, at least within the context of the work place, managers have less opportunity for establishing satisfying social relationships than do non-managers. Conflicts of interest and differing levels of responsibility and decision making are two of the main reasons for this. This is simply another reason that it is important for managers to have a clear understanding of their existing and desired level of social support and any discrepancies between them. Simply put, the network of people whom you consider supportive in large and small ways, and to whom you can turn, can be a very important factor in handling stress and in maintaining good health.

The following exercise is designed to help you better understand your own sources of friendship and support. Several discussion questions follow the exercise.

Think about the groups of people listed in Table 6.2 overleaf and indicate how you feel about each of them as a source of support by placing the appropriate number next to each item.

3 = A highly important source of support for me.
2 = A fairly important source of support for me.
1 = Not an important source of support for me.

Table 6.2 *Social support*

1 Immediate family members	
2 Relatives	
3 In-laws	
4 Friends	
5 Experts (doctors, plumbers, lawyers, teachers etc.)	
6 Helpers (people who are happy to provide direct help)	
7 Reference groups (race, gender, professional groups etc.)	
8 Challengers (people who encourage you to excel and grow)	
9 Access providers (people who can link you to people or things you need or want)	
10 Energizers (people who excite or stimulate you)	
11 Respecters (people who make you feel good about yourself)	
12 Co-workers	
13 Professional colleagues	
14 Casual acquaintances	
15 Recreational participants (people you share leisure activities with)	
16 Other (name)	
17 Other (name)	

 Do any of your present relationships with people in your social support network need to be improved? In what ways?

Are any of your present relationships with people in your social support network beyond saving – needing to be ended? Why is this?

Are there any important social qualities (for example, expertise, challenge and so on) missing from your present social support network? What are they?

Do you know people who would be valuable additions to your social support network? Who are they?

Are the members of your present social support network aware that you consider them to be your supporters?

As you reflect on your present social support network and any needs for change, what is the one step that you can take in the near future to improve the quality of your social support?

One of the best ways to get through a stressful period is to turn to a social support network for counselling, encouragement, direct assistance or expertise. Unfortunately, many people have a tendency to withdraw from their support networks during times of stress, rather than actively using them for support.

Frequently, it also happens that one or more key members of your social support network engage in the same high-risk lifestyles (for example, smoking) that you yourself may wish to change. They may not be very supportive of the changes you want to make and may even actively oppose you! You may therefore find it necessary to do some specific negotiating or bargaining. As your outlooks and habits change, you may even find it necessary to change or perhaps end some relationships and form new, more supportive ones.

In reviewing your responses to the social support questionnaire, try to answer the following questions.

How do you feel about the size of your present social support network?

_____

_____

How do you feel about the diversity of your present social support network?

_____

_____

Are you relying on a very few people for too many different social support qualities?

_____

_____

Do you need/want to do anything about this?

_____

_____

Diet

Diet is one of the staying healthy factors over which you as a manager have most control. While you are not in a position to supervise the preparation of each meal, managers generally have more discretion over when and under what conditions they eat and, if you are a successful manager, you have probably developed good planning skills. If you are unaware of basic diet and nutrition facts, then this section should be especially useful to you. Read it carefully and refer to it as you develop better dietary behaviour.

The section begins with a brief review of your dietary behaviour and personal nutrition. Nutrition is a central factor in maintaining and improving health. To check your current attitudes to nutrition and diet, place write one of the following ratings beside each item in the list in Table 6.3 overleaf.

0 = Almost never
1 = Rarely
2 = Sometimes
3 = Often
4 = Nearly always

Review your responses. Compare your actual food-intake pattern with your ideal dietary behaviour. Are there discrepancies? Are you concerned about them? What steps might you take to reduce the discrepancies?

If you have any health issues that could be related to an unhealthy diet, have you consulted a doctor about these? Does diet and food intake have any negative impact on your performance as a manager? If so, now may be the time to deal with such issues.

There are many excellent books and materials about diet and nutrition available. If, after reviewing your responses to the diet items, you have concerns, these could be helpful sources of information. Knowing how to eat healthily doesn't require a degree in nutrition. Maintaining a healthy diet once you are informed may be another question, but there are ways of controlling eating behaviour. Two useful techniques are planning and recording your food intake. Plan a menu for each week. Note the items you plan to eat for each meal and check their nutritional values to see if they make good health sense. Then record daily what you actually consume. If you exceed your limits one day, cut back the next. Try to end each week in a balanced position.

Table 6.3 *Dietary questionnaire*

1 Cooking and meal times are a source of happiness	
2 Time is taken for leisurely meals	
3 Worries, anger and troubles are set aside during meal times	
4 I read and educate myself about good nutrition	
5 I consider my own nutritional needs and how to meet them	
6 I eat a well-balanced breakfast	
7 I eat three or more well-balanced meals per day	
8 I eat fruits and vegetables daily	
9 I eat whole grains every day	
10 I include enough fibre in my diet	
11 I keep my weight within 10lb of my ideal	
12 I am aware of the effect of blood sugar levels on energy and performance	
13 I am aware of the effects of food allergies or sensitivities on my energy level and performance	
14 I chew my food thoroughly	
15 I consider my needs for nutritional supplements	
16 I am open to changing my eating habits when I gain new information	
17 I avoid processed goods and packaged meals	
18 I maintain good eating habits when upset or bored	
19 I use food labels to monitor fat and calorie counts	
20 I average no more than two alcoholic drinks per day	
21 I do not smoke tobacco	

Vitamins and minerals

A good eating plan should incorporate a variety of foods including cereals, grains, fruits, vegetables, dairy products and protein. It should minimize refined white flour, sugar, salt, saturated fats, artificial additives and alcohol. It also means getting the right balance of vitamins and minerals.

A great deal of attention has been focused on the role of minerals in keeping us healthy. Although they are readily available in most diets, they are used up very rapidly in times of stress. Fresh vegetables and fruits are the best source of many minerals, as long as they are not over-cooked or soaked for too long. Meat provides a good source of iron, as do green vegetables. Calcium is readily available from dairy products. Refined foods, like bleached flour, often have reduced mineral content, so try to eat whole foods and fresh produce.

Tables 6.4 and 6.5 give details of the various vitamins and minerals we need and how they can be obtained.

You should note that many vitamins and minerals cannot work alone; they need to work in combination to be effective. Thus it is even more important to make sure that you eat a balanced diet. Too much of one vitamin or mineral – salt, for instance – can be as harmful as too little.

Table 6.4 *Minerals*

MINERAL	PROPERTIES
Zinc	Essential for mental alertness and the speedy healing of wounds. Lack of zinc can cause loss of appetite, brittle nails and loss or dimunition of taste and smell.
Food sources	*Most red meat, whole grain cereals, wheatgerm, squash and pumpkin seeds, peanut butter, turkey, lentils, plain yoghurt, eggs*
Calcium	Needed for healthy teeth and bones and to maintain muscles and nerves. Vitamin D is needed to help absorb calcium.
Food sources	*Dairy products*
Iron	Necessary for healthy blood, as it is needed to form the red blood cells that transport oxygen around the body. Constant tiredness and itchy skin are symptoms of iron deficiency.
Food sources	*Most red meat, most green vegetables, pork, kidney beans, lentils, spinach, prune juice*
Magnesium	Prolonged illness or heavy drinking can reduce magnesium levels, so make sure that you get enough magnesium at times of stress. It helps fight depression and can also help to relieve stomach upsets.
Food sources	*Most foods, especially nuts, cereal grains, legumes, cocoa, dark green vegetables*

Table 6.5 *Vitamins*

Vitamin	Best food sources	Function	Deficiency signs	Adults' minimum daily need	Lack may be caused by
A – Retinol	Fish liver oil, oily fish, liver, kidney, dairy foods, margarine, green vegetables, yellow fruit, carrots	Essential for growth health of eyes, structure and health of skin	Low resistance to infection, night blindness, catarrhal and bronchial infections, skin complaints	2500 iu	Frying, canning, long-term treatment with liquid paraffin, some drugs
B1 – Thiamin	Yeast, wheatgerm, meat, nuts, beans, whole grain foods, pulses, seafood	Essential for growth, conversion of carbohydrates into energy, health of nerves, muscles	Nervous disorders, easy exhaustion, depression, poor digestion	1mg	Alcohol, tobacco smoke, over-cooking, sulphur dioxide, high-sugar diet
B2 – Riboflavin	Yeast, wheatgerm, meat, soya beans, eggs, vegetables, cheese	Essential for growth, health of skin, mouth, eyes, general well-being	Dry hair and skin, mouth sores, nervousness, lack of stamina	17mg	Contraceptive pill
Pantothenic acid	Yeast, liver, wholemeal bread, brown rice, eggs, nuts	Health of skin and hair, including hair growth, needed for all tissue growth	Dry hair and skin	5–10mg	Heating above boiling, absent from many refined foods
B6 – Pyrodoxine	Yeast, wheatgerm, meat, fish, wholemeal products, milk, cabbage, nuts	Essential for body's use of protein, health of skin, nerves and muscles	Irritability, depression, skin eruptions, insomnia, muscle cramps	1mg – women taking oral contraceptives need much more	Loss in cooking and freezing, contraceptive pill, alcohol, smoking
B12 – Cyanocabalamin	Liver and meat, eggs	Health of nerves, blood and skin, body's use of protein, growth	Anaemia, tiredness, skin disorders	15mcg	Veganism, malabsorption, contraceptive pill, many drugs, smoking
Biotin – B group	Liver, kidney, wheatgerm, bran, oats, eggs, nuts	Probably essential for healthy skin, nerves and muscles	Falling hair, eczema	1mcg	
Choline and inositol – B group	Eggs, liver, yeast, offal, wheatgerm, oats, nuts	Essential for functioning of liver, prevent build-up of fats in body	Liver disorders, reduced alcohol tolerance	10mg of each	Deficiency unlikely
Folic acid – B group	Offal meats, green vegetables, yeast, wheatgerm, soya flour	Essential for all growth, healthy blood, fertility	Anaemia, weakness, depression, diarrhoea	0.5mg	Pregnancy, many drugs including contraceptive pill, canning, prolonged cooking
Niacin nicotonic acid – B group	Meat, fish, whole grain products, peanuts	Essential for growth, health of skin, digestion of carbohydrates, nervous system	Skin disorders, nervous and intestinal upset, headaches, insomnia	15–18 mg	Missing from over-refined foods
C – Ascorbic acid	Citrus fruit, other fruit, raw vegetables green and red peppers, potatoes	Essential to health of cells, blood vessels, gums and teeth, healing of wounds	Sore gums, low resistance to infection, slow healing, painful joints	30mg	Many drugs including aspirin, corticosteroids, contraceptive pill, over-cooking, refining of foods, alcohol, smoking
D – Calciferol	Fish liver oils, sunshine on skin, butter and margarine, eggs	Formation of bones and teeth, needed for calcium and phosphorus use	Retarded growth, crooked bones (rickets), tooth decay, weak muscles	100iu	Lack of sunshine or dairy foods, and other factors as for Vitamin A
E – Tocopherol	Vegetable oils, wheatgerm, wholemeal bread, egg yolks, green vegetables, nuts	Known to be essential but function not fully understood. Needed by animals for fertility and muscle health	Muscular disorders, infertility and nervous disorders in animals	Not certain, but estimated at 10mg	Air, light, deep freezing, frying, refining of vegetable oils, contraceptive pill
K	Green vegetables, soya beans, liver, oils	Essential for blood clotting	Prolonged bleeding from cuts or sores	Unknown	Similar to E, antibiotics, liquid paraffin, malabsorption, diseases

The tendency in all affluent countries is to consume large amounts of animal protein and fats in the form of meats and cheeses. This is combined with a tendency to consume fewer complex carbohydrates like whole grains, nuts, fruits and vegetables, nutritious sources of proteins, vitamins, minerals and fats. In addition, in a busy society, commercially prepared meals and convenience foods have taken precedence over whole, unprocessed foods, as we often assume that it takes too much time to prepare good meals.

One of the most important things to consider relative to stress and health is the need for B and C vitamins. These are burned up very quickly under stress and are in somewhat short supply in most diets. A further complication is that sugar, caffeine, tobacco and alcohol all diminish the availability of these necessary nutrients. If you consume a lot of any of these substances during periods of stress, you are adding to your problem. Be sure to eat plenty of fibre and complex carbohydrates and avoid sugar, caffeine, tobacco and alcohol at these times.

Jot down any discrepancies (that is, too much or too little of certain foods) evident in your current eating pattern.

For most people the major problems connected with eating habits are as follows.

Excessive fat
Most people eat far too much fat – on average about 45 per cent of our calories. Nutritionists suggest that we should consume only 25 per cent of our calories in the form of fats. Excessive fat consumption is implicated as a risk factor for cardiovascular diseases and several forms of cancer.

Can you think of any way in which you could reduce your intake of fats?

We suggest that you can control your intake of saturated fats by eating less red meat, high-fat cheeses, dairy produce, palm and coconut oils. Reduce the amount of fried foods that you eat, together with cakes, pastries and chocolate. Instead eat chicken, fish or vegetables cooked with little or no fat. Skimmed or semi-skimmed milk is a good idea, as it still contains calcium but has a reduced fat content. Also look out for reduced-fat cheeses, yoghurts and so on. If you use oils and spreads, try to use the ones marked 'high in polyunsaturates'. Instead of eating cakes at the end of a meal, eat fruit instead. There are many exotic varieties of fruit available, so your choice need not be limited to apples and oranges! By cutting down like this you will not only do your heart good, you will probably lose weight too, especially if you also follow the next set of guidelines.

Excessive sugar

The annual per capita consumption of sugar is increasing and now exceeds125lb per year. Some of this sugar occurs naturally in foods, but about two-thirds is added by food manufacturers. For example, a 12oz can of soft drink contains between 9 and 11 teaspoons of sugar. In addition to adding many unnecessary calories to the diet, sugar uses up a great many B and C vitamins and causes triglycerides, a cardiovascular risk factor, to increase. Most people who consume sugary foods eat something containing sugar every two to three hours to maintain their energy.

You can cut down your sugar intake by avoiding sweets, cakes and soft drinks. Many processed foods contain large amounts of sugar, so steer clear of these or read the ingredients carefully to check sugar levels.

Excessive salt

We need small amounts of sodium on a daily basis, but most people eat five to ten times more than they need. This causes fluid retention on the cellular level, forcing the heart to work harder to feed the cells. As a result, blood pressure is increased. Try reducing salt levels in your food or use a salt substitute. There are hundreds of herbs and spices that can add flavour to food but don't have the harmful effects of too much salt.

Insufficient fibre

Most people get only about 15 per cent of their calories from fruits, vegetables, grains, cereals and nuts – plant foods.

Nutritionists tell us that we should be getting four times that much! All the fibre in our diet has to come from plant sources. Fibre is necessary for gastrointestinal health (it probably prevents colon cancer) and can significantly reduce cholesterol. Many kinds of fibre are necessary and you can only get the full range by eating a broad variety of plant food. This is also the main source of vitamins B and C, which are essential in greater quantities during times of stress. The action here is obvious – cut down on meats, cheeses, fatty and sugary food and eat more plant foods!

Excessive caffeine

There is about 100mg of caffeine in a cup of coffee. People who consume over 300mg of caffeine a day have increasingly irregular heart beats. It also stimulates the adrenal glands to produce stress hormones. Caffeine is antagonistic to B and C vitamins and reduces the blood sugar, stimulating feelings of hunger. Instead of drinking coffee, try to drink more fruit juices, herbal teas or mineral waters.

Try keeping an eating diary for a week, recording everything you eat and drink. Then review your entries and estimate your total calorie consumption.

Most people want to lose several pounds in weight. If this is true of you, where in your diet can you cut fat and sugar calories? If you use the above nutritional problems as a guide, analyse your food diary and note where the excesses and deficiencies are, you will probably notice some clear patterns in your eating habits regarding fat, sugar, salt, caffeine and complex carbohydrate consumption. Necessary changes are likely to be obvious, so write down some possible alterations to your diet.

Physical exercise

Regular physical exercise is an excellent way to reduce stress and to improve circulation, metabolism, flexibility and heart and lung efficiency. It can also enhance mood and creativity and even alleviate some kinds of pain. Developing and following a regular exercise pattern is one of the most important health-maintenance behaviours available and can be especially important for managers. The responsibilities entailed in management are often stress related and many managerial positions involve very limited amounts of physical activity.

In addition to a regular exercise schedule, managers can often increase their physical activity by being aware of subtle opportunities for muscle use. This typically requires avoiding, whenever possible, convenient options such as lifts, escalators and short trips by car or taxi. Using these modern conveniences does save time, but if you are looking for ready-made opportunities for exercise the option is worth considering.

The checklist in Table 6.6 is designed to help you review your exercise pattern. Indicate your present exercise habits by placing the appropriate numbers beside each item on the list.

0 = Never
1 = Rarely
2 = Sometimes
3 = Often
4 = Usually
NA = Not applicable

These items do not lend themselves to scoring, but you can review your ratings to get a general impression of your exercise behaviour and awareness. You can also use the results as you read the following reflections on exercise.

Our bodies are designed to be used vigorously and we have become sedentary. It has been estimated that in the middle of the nineteenth century at least one-third of the power used in work was supplied by human muscle. Today that figure is less than half of 1 per cent. Without sufficient exercise, we can tol-

Table 6.6 *Exercise checklist*

1 I take some form of recreational exercise (slow walking, golf, bowling, gardening etc.) at least once a week	
2 I take some form of stretching exercise at least three times a week	
3 I raise my heart rate aerobically, so that I am slightly out of breath two or three times a week (e.g. swimming, exercise class) or I jog 3 miles in 18 minutes or walk 3 miles in 42 minutes (or the equivalent) three to five times a week	
4 I find exercise to be an enjoyable and rewarding pastime	
5 I am aware of the effects of my present exercise level on my resting pulse rate	
6 I am aware of the effects of my present exercise level on my blood pressure	
7 I am aware of the effects of my present exercise level on my digestion and metabolism	
8 I am aware of the effects of my present exercise level on my energy level and performance	
9 I am aware of the effects of my present exercise level on my body fat percentage	
10 I am aware of the effects of my present exercise level on my general mood and level of self-confidence	
11 I use stairs instead of lifts and escalators	
12 I walk short distances instead of taking a taxi	
13 I walk short distances instead of using the car	
14 I use hand tools instead of powered devices in the garden	
15 I walk unnecessarily at home and at work	

erate fewer calories, are less sharp mentally and less coordinated physically. Our cardiovascular and respiratory systems become sluggish and we break down faster.

Stress causes the endocrine glands to produce 'fight or flight' hormones which, if not expressed, accumulate and drive our strain scores up. Regular, vigorous exercise (20 to 30 minutes per session, three to five sessions a week of sustained exertion) burns up these hormones and helps your body chemistry return to normal. Exercising at this level has benefits that help to restore or protect health. In many cases regular, vigorous exercise causes a lowering of blood pressure. It also lowers the resting pulse rate, indicating that the heart and lungs have become stronger. Exercise normalizes metabolism and improves digestion, which are added aids to weight control. Furthermore, exercise can burn up excess body fat and improve both your energy level and your performance.

On the psychological level, exercise can improve mood, confidence and creativity by triggering the production of brain chemicals called endorphins. These endorphins also serve to alleviate pain and withdrawal symptoms, making exercise an excellent support for people who are giving up tobacco or alcohol!

There are many good books on exercise that may be of assistance to you. All of them suggest that you have a thorough physical check-up before beginning and that you develop your level of fitness gradually. You should select an exercise programme (running, swimming, walking, bicycling or stationary exercise equipment) that you enjoy. It doesn't have to be drudgery or boring – it should be stimulating and enjoyable. Try several things to find what suits you best.

What goal are you setting for yourself to improve your physical fitness?

Alcohol consumption

Alcohol, when abused, is a well-known risk factor. Because managers are in a relatively high-stress group and socializing at work can often involve drinking, they need to be especially aware of this health risk. There is, as you may know, lack of agreement among scientists as to whether excessive alcohol consumption is a disease or simply a bad habit. Putting that argument aside, if you have any concerns about the amount of alcohol you consume, what it may signify and whether it is considered a problem by society, we recommend that you complete the set of questions in Table 6.7 and do the brief follow-up exercise.

For each question that is true about you, place a tick in the 'yes' column on the line beside the question. For each question that is not true of you, place a tick in the 'no' column.

Table 6.7 *Alcohol questionnaire*

	Yes	No
1 Do you feel that you are a normal drinker? (Yes, if you do not use alcohol at present.)		
2 Have you ever woken up the morning after drinking the night before and found that you could not remember a part of the evening?		
3 Does your partner or family ever worry or complain about your drinking?		
4 Can you stop drinking without a struggle after one or two drinks?		
5 Do you ever feel bad about your drinking?		
6 Do your friends and relatives think you are a normal drinker?		
7 Do you ever try to limit your drinking to certain times of day or certain places?		
8 Are you always able to stop drinking when you want to?		
9 Have you ever attended a meeting of Alcoholics Anonymous?		
10 Have you ever got into fights when drinking?		
11 Has drinking ever created problems between you and your family?		
12 Has your partner, or another family member, ever gone to anyone for help about your drinking?		
13 Have you ever lost friends because of your drinking?		
14 Have you ever got into trouble at work because of your drinking?		
15 Have you ever lost a job because of your drinking?		
16 Have you ever neglected your obligations, your family or your work for two or more days in a row because you were drinking?		
17 Do you drink before noon fairly often?		
18 Have you ever been told that you have liver trouble? Cirrhosis?		
19 Have you ever had delirium tremens (DTs), severe shaking, heard voices or seen things that weren't there after heavy drinking?		
20 Have you ever gone to anyone for help about your drinking?		
21 Have you ever been in a hospital because of your drinking?		
22 Have you ever been a patient in a psychiatric hospital or on a psychiatric ward of a general hospital where drinking was part of the problem?		
23 Have you ever been seen at a psychiatric or mental health clinic, or gone to a doctor, social worker or clergyman for help with an emotional problem in which drinking played a part?		
24 Have you ever been arrested, even for a few hours, because of drunken behaviour?		
25 Have you ever been arrested for drunk driving or driving after drinking?		

Reprinted with permission of the AAA Foundation for Traffic Safety from the *DWI Counselling Manual*.

The questionnaire is scored by giving yourself one point for each 'key' response. Your total score is the number of key responses.

The key responses are as follows:

1 No	6 No	11 Yes	16 Yes	21 Yes
2 Yes	7 Yes	12 Yes	17 Yes	22 Yes
3 Yes	8 No	13 Yes	18 Yes	23 Yes
4 No	9 Yes	14 Yes	19 Yes	24 Yes
5 Yes	10 Yes	15 Yes	20 Yes	25 Yes

Drinking habits score:
0–3 = No drinking problem
4–6 = Potential alcohol problem
7–25 = It would appear that you already have an alcohol problem. We suggest that you seek some professional help or advice
as soon as possible.

For many people, one or two drinks a day may be relaxing and a
means to enliven social interactions. When taken in larger
amounts, alcohol has a depressant effect and directly affects behaviour, reflexes, self-control and judgement. Thus the risk of an acci-
dent is dramatically increased as more alcohol is consumed.
Alcohol is the single biggest factor in all road deaths and injuries.

The risks of serious illnesses are also increased by continual
heavy drinking – cirrhosis, cancer, pneumonia and heart disease
are all made more likely. The abuse of alcohol is also implicated
in a high percentage of cases of family problems and financial
and job difficulties.

It is estimated that between 8 and 14 million working days
are lost in the UK each year because of heavy drinking. At work,
alcohol and drug abusers are three times as likely to be late, use
three times as much sick leave, make five times as many com-
pensation claims and have more than three times as many acci-
dents as non-users.

**Can you think of any ways in which you can cut down your
alcohol intake?**

_____

_____

We suggest that you start by keeping a 'drinking diary' to measure the average amount of alcohol you consume each week. Also include the situation – for example, at the pub with friends, lunchtime with family. You may find that there are times when you drink more than is necessary, perhaps after a hard day at work or when you are with a particular friend. Make a note of these and watch for similar situations in the future.

If you scored in the risk area of the alcohol-assessment questionnaire, you are encouraged to seek immediate assistance. Alcohol abuse is an extremely risky activity and should not be treated lightly.

Other behaviour and concerns

This final section provides information on several other common health concerns. Many books and materials are available covering these issues. As with all sections in this chapter, you are urged to pursue further information about any that worry you particularly.

Type A behaviour
You probably know about type A and type B behaviour. Some people believe that managers are especially prone to type A behaviour. This brief overview should answer some basic questions about this. It begins with a brief questionnaire (Table 6.8). The questionnaire is not scientifically designed and is not intended as a diagnostic instrument. It should, however, provide you with a general guide to considering the implications of your behaviour.

Indicate how often each of the behaviours in the list is true of you by placing the appropriate number on the lines beside the items. When you are finished, total your marks and compare them with the points indicator following the items.

0 = Never
1 = Very rarely
2 = Rarely
3 = Sometimes
4 = Often
5 = Always

Table 6.8 *Behavioural habits questionnaire*

1	I schedule more and more activities into less and less time	
2	I become irritated and impatient when I am delayed or made to wait	
3	I become impatient when watching others do things that I can do faster or better	
4	I have difficulty sitting and doing nothing	
5	I want to win every game I play, even when I am playing with children	
6	I work harder to get things done than most of my colleagues	
7	I become angry when I see inefficiency or carelessness in others	
8	I am easily aggravated or frustrated by events which, a short while later, seem trivial	
9	I keep myself too busy to 'stop and smell the flowers'	
	TOTAL SCORE	

Points indicator:

40–50 = strong type A = high risk
35–39 = type A = moderate risk
20–34 = type AB = moderate risk
15–19 = type B = low risk
0–14 = strong type B = low risk

This questionnaire on behavioural habits measures type A behaviour tendencies. Look at Table 6.9 for a brief outline of what are classed as type A and type B behaviours.

Type As are also more likely to smoke, take little exercise and have fewer holidays than type Bs – making their lifestyle even more stressful!

Which type are you?

_____

_____

Table 6.9 *Behavioural types*

Type A	Type B
Competitive	Relaxed
Achiever	Easy-going
Fast worker	Seldom impatient
Aggressive	Takes time to enjoy pursuits outside work
Impatient	Works steadily
Restless	Not easily irritated
Hyper-alert	Seldom short of time
Explosive speech	Moves and speaks slowly
Frequently feels under pressure	Not preoccupied with achievement

Do you know someone who is the opposite to you?

Type A behaviour is now considered to be a primary risk factor for coronary heart disease. Those who regularly display excessive amounts of anger, impatience, irritability and urgency, especially in situations in which these reactions can do nothing to resolve matters, are much more likely to experience a heart attack than those who permit themselves to take things as they come and who only 'engage' in situations that can be influenced. For example, losing one's temper in traffic jams or in queues at the supermarket does nothing to speed up traffic or the queue, but may do a great deal to speed up the onset of heart disease.

Type A behaviour causes your body to enter into and chronically maintain the 'fight or flight' response. This means that total cholesterol, pulse rate and blood pressure are likely to become chronically elevated. If your behavioural habits score is in the risk area, it may be important for you to learn that you can still work hard and not be irritable or aggressive at the same time. It is not the hard work or high productivity that is thought to be the risk factor.

One of the most important things for you to do if your score is in the risk area is to learn to relax. Relaxation reverses the signals that the brain is sending out from 'fight or flight' to 'relaxation'. Many people who score high on measures of type A behaviour are very resistant to the idea of relaxation, stating that it is 'such a waste of time – you don't accomplish anything!'. So this suggestion of learning to relax is likely to stir up some resistance, especially if you really need to learn how! You may want to review again the relaxation techniques described in Chapter 3. In reality, learning to relax effectively takes a great deal of discipline and determination, so approach it as a challenge.

Here are some relaxing activities:

◆ going for a leisurely walk
◆ listening to your favourite music
◆ having a long, hot bath.

Can you add some more?

_____

_____

Another suggested way to control type A impulses is to practise patience in situations that are out of your control and in which you often lose your temper and become impatient. The next time you find yourself in a traffic jam or at the end of a long queue, say to yourself 'I choose to remain calm' and see what happens. If you repeat this on each such occasion, you will soon find that you are not losing your temper nearly so often.

Additional factors
This section considers various other factors that have a bearing on health and stress management. If you do not know your blood pressure or cholesterol level, omit these questions

and make a note to ask your doctor next time you have a check-up.

Do you use tobacco?

_____

_____

If you once smoked and have stopped, how long have you been a non-smoker?

_____

_____

What is your blood pressure?

_____

_____

What is your cholesterol level?

_____

_____

Smoking

Smoking is the undisputed number one health risk in our lives today. A smoker is two or three times more likely to have a heart attack and at least eight times more likely to develop lung cancer, coronary heart disease and chronic obstructive lung diseases.

In order to stop smoking, an individual must first clearly choose to be a non-smoker. Once this fundamental choice is made, there are a number of methods available to support them in stopping.

A balanced diet and plenty of exercise are especially important during the period of stopping. While it takes about 10 years from the time one stops smoking to achieve the health-risk status of someone who has never smoked, most of the risk has disappeared within the first two years (see Table 6.10).

Table 6.10 *Smoking risk*

Never smoked or stopped over 15 years ago	Stopped smoking 5–15 years ago	Stopped smoking less than 5 years ago	Current smoker less than 20 cigarettes a day	Current smoker 20–40 cigarettes a day	Current smoker more than 40 cigarettes a day
Low risk	Moderate risk		High risk		

Blood pressure

Your blood pressure is the pressure which is exerted on the walls of your blood vessels by your blood as it circulates through your body. The first (larger) number is called the 'systolic' pressure and is a measure of the pressure in your arteries when your heart contracts. The second (smaller) number is called the 'diastolic' pressure and is a measure of the pressure in your arteries between heart beats. A blood pressure of 110/70 is read '110 over 70' and means 110mm in a mercury column of systolic pressure and 70mm in a mercury column of diastolic pressure.

Blood pressure greater than 140/90 is considered to be high and steps should be taken to bring these figures down. In general, the lower the numbers the lower the risks of developing heart disease, atherosclerosis, a stroke or kidney damage. It makes sense that if your arteries are subjected to prolonged high pressures the risks are increased. The heart has to work harder, the artery walls tend to harden and narrow and the tiny arteries in the kidneys are damaged, reducing that organ's ability to clear wastes from the blood.

High blood pressure does not have any symptoms associated with it, so many people who have it are unaware of their condition. In addition, since people with high blood pressure don't have further symptoms and because the medications often create side effects, many people do not continue taking the medication prescribed to reduce blood pressure. Prevention is of course the best strategy and if blood pressure does become chronically elevated, there are several natural lifestyle-related steps, outlined below, that can be used.

If your blood pressure is raised, there are several things you can do. First, you should get your blood pressure checked regularly. You can even take it yourself very easily. There are a number of accurate and inexpensive measurement devices available

Table 6.11 *Blood pressure risk*

Systolic reading						
Under 110	110	120	130	140	150	over 150
		Low risk		Moderate risk		High risk

Diastolic reading						
Under 70	70	80	90	100	110	over 110
		Low risk		Moderate risk		High risk

at any well-equipped local pharmacy. Second, if you are over-weight, you may be able to reduce your blood pressure to 120/80 or below simply by losing weight.. You should also watch your salt consumption very carefully – remember that most of the salt we eat comes from processed foods. Also be sure to get plenty of exercise and build some kind of relaxation routine into your life, as both of these activities help to lower blood pressure. Finally, if these do not work, your doctor can prescribe medications that will lower your blood pressure.

Cholesterol and triglycerides
Cholesterol and triglycerides are fatty substances produced in the body and carried in the bloodstream. They are necessary for normal cell functioning. When they are produced in excessive quantities they increase one's risk of developing heart disease. Cholesterol is increased by eating foods that contain saturated fats and/or cholesterol. Triglycerides are increased by eating foods high in sugar and by alcohol consumption. As with most biological processes, some people are more inclined to produce high levels of these substances than others.

Triglyceride levels respond very quickly, within a matter of days, to changes in sugar and alcohol intake. Many people have found that if their triglyceride level is raised it is difficult to alter their cholesterol level through dietary changes.

Cholesterol's role in cardiovascular health has been exten-sively documented. It is known that cholesterol levels are directly related to heart disease, since excess cholesterol in the blood is deposited on the walls of arteries. When an artery that feeds the heart muscle becomes clogged, the result is a heart attack; when this happens in the brain, the result is a stroke.

Cholesterol travels through the bloodstream in protein 'packets' called lipoproteins. Some of these packets are low density (LDL) and some are high density (HDL). The LDL form transports cholesterol from the liver to the rest of the body. This is the form of cholesterol that adheres to artery walls and causes problems. The HDL form transports cholesterol out of the body and is now thought to be protective against heart attacks.

As a result of these different functions of the two forms of cholesterol, the current medical advice is that people reduce total and LDL cholesterol and increase HDL cholesterol. A diet low in saturated fats and cholesterol-rich foods will help to lower total and LDL cholesterol levels in most people. A regular, vigorous exercise programme will increase the HDL proportion. Thus a combination of a low-fat/low-cholesterol diet and a regular exercise programme are strongly suggested for protection or to reduce risk.

 Do you know what foods are high in saturated fats and cholesterol?

Saturated fats are those that are firm at room temperature (for example, butter, cheese, beef and pork fat, coconut and palm oils). A very common operation used by food manufacturers called 'hydrogenation' causes unsaturated oils – for example, corn oil – to become saturated. Cholesterol is concentrated in red meats, especially organ meats like kidney and liver, and in dairy products.

In addition to reducing the intake of foods rich in cholesterol and saturated fats, eating one-third of a cup of oat bran or corn bran each day is a good, natural way to reduce cholesterol. A number of good books about cholesterol management are now available and can be referred to for more detailed information.

Table 6.12 *Cholesterol risk*

Total cholesterol

170 and below	190	210	230	250	270 and above
Low risk			Moderate risk		High risk

Staying current

That's a very large dose of mind/body information! What makes it even more interesting is that what is true today about your mind/body health status may not be true tomorrow. Life is dynamic, thankfully, and thus as conditions change, our health changes.

The obvious message here is to maintain an ongoing awareness of your mind/body health status. One way to do this is to make a note in your diary or planner to review the health-assessment items in this chapter. Such a review can be viewed as an extension of any regular medical examinations that you might have. Devote a few minutes every six months or so to assessing yourself. The results will be useful for your ongoing health-management plan. We look at developing such a plan in Chapter 7.

7

Health Planning

As you well know by now, staying healthy doesn't just happen and, as a manager, you understand that most good things are more likely to happen as a result of planning. Whether your management style is characterized by objectives, by flowcharts and diagrams, by 'walking the floor' or by something else, it is likely that you don't rely on luck for results. So why should staying healthy be any different? Certainly, as in any part of life, luck and happenstance play their parts but, as we have tried to demonstrate in this book, there are many ways in which you can influence your personal health and perhaps that of your staff.

This short chapter is in essence a motivational message. It ends by encouraging you to generate health goals for the year and to project criteria for knowing how successful you have been in achieving these. It is about taking charge of your health and making good health happen.

Here is one suggestion for doing that. If you believe that 'managing is managing is managing', then wouldn't it be a good idea to identify the management techniques and procedures that work best for you as a manager (and those that you have observed working well for other managers) and apply them to your personal health objectives and goals?

'Oh, that's different,' you're probably thinking. Is it?

It is not unusual for successful managers and other professionals to encounter difficulty in their personal lives. A question we often ask these people is how is it that that are so adept at managing their work and professional life, but such an embarrassment to themselves when it comes to personal and family management. As you might have guessed, this question isn't always well received. What we often get in response is something to the effect 'Well, my working life is different. It's much more clear cut, more objective, with fewer variables and surprises to deal with than there are in my personal life.'

A technique we use to demonstrate that they are talking nonsense is to go through the following exercise. We reach into the file and pull out a folder containing details of a relatively complex, but not atypical, personal problem. (It's not an actual case, of course, but a good hypothetical one.) We ask managers to read the case summary and then describe what advice they might offer to the person in the case. You will probably not be surprised to learn that advice comes quickly and with little hesitation. It amounts to a summary of their favourite problem-solving strategies in management.

You might be ahead of us; the next step is fairly obvious. It is simply to suggest that successful managers follow their own advice. Apply your management and problem-solving skills to your own health issues. Some of the variables are different but there are also many similarities. Understand what you have to work with, identify needed resources, plan and schedule. During the process of establishing a health-management plan, it may be useful to reward yourself along the way. Sometimes success in meeting your goals is sufficient. In other words, use your favourite management techniques on your most valuable personal asset – your health. Let the planning begin.

People don't always begin to do something with the kind of information contained in this book immediately. They tend never to get around to making the changes they might need to make in order to reduce the risks to their health and improve their performance and sense of well-being. You really need to take immediate action to get the ball rolling.

New Year's resolutions and other good intentions often don't reach a successful conclusion. There are generally two reasons for this. The first is that people often try to change too much at once, become overwhelmed, then give up. The second is that the context in which the resolution is made (for example, New Year's Eve in high spirits) differs from that in which the changes need to be made (for example, the first Monday of the new year in the company dining room).

Can you recall any occasions in the last six months when you have made resolutions or promises that you have not kept?

_____

_____

Think carefully for a few minutes and then write down your thoughts about why your good resolutions failed.

_____

_____

You will be more successful if you avoid attempting wholesale changes and instead think about making one small change at a time. For example, start by not eating sweets during the week or only smoking after two o'clock in the afternoon. That way, you will have successes rather than the failures that might come with attempting to do too much, too soon. With this approach, you will find that you are able to accumulate gradually significant changes in a way that does not cause you to feel you are making great sacrifices. Those who are most successful at making personal changes are the people who accept that they are life-long learners and make changes one step at a time.

If you want to be successful in taking charge of your life, you must affirm that you are in charge and you must consciously choose to be healthy. When you do this, you vastly improve your chances of success. If you are unable to say to yourself, in a strong and convincing way, that you are in charge and that you choose to be healthy, it probably won't matter much what changes you attempt, since it is unlikely that you will realize the benefits of those changes.

Use the following outline to identify the possible projects that you think you should eventually undertake and then decide what your first project will be. The first project should be something that is manageable, something that you are confident you can accomplish and something that you are committed to undertaking right away. Early successes create momentum that you can use when you come to the more difficult changes.

In reviewing all the information you have gathered about your stress levels and about your attitudes, lifestyle and habits, complete the following personal action plan.

Action plan

The things that I am doing well and want to maintain are:

The things that I need to stop doing or do less are:

The things that I need to start doing or do more of are:

Now list these in order of priority:

Which one of these shall I undertake now or do in the very near future?

The first steps I plan to take are:

_____

_____

The factors that will hinder me are:

_____

_____

The factors that will help me are:

_____

_____

The implications of not acting are:

_____

_____

The support that I will need is:

_____

_____

The agreements that I need to make are:

_____

_____

I will know that I have been successful when:

_____

_____

Congratulations! You have now completed quite a thorough review of your stress levels and those lifestyle factors that are most important for protecting and improving your health and performance. As you continue to follow up on what you have learnt from your efforts, you will find that you are increasingly satisfied with your life and that you are gaining more and more control over your responses to events.

Suggested additional reading

Adair, John (1985) *Effective Decision Making*, Pan.

Allen, Madelyn Burley (1995) *Managing Assertively: How to Improve Your People Skills – A Self-teaching Guide*, Wiley.

Body Shop, The (1998) *Mind, Body and Soul: The Body Shop Book of Wellbeing*, Ebury Press.

Bower, Sharon Anthony and Bower, Gordon H. (1991) *Asserting Yourself*, Addison Wesley.

Carlson, Richard (1999) *Don't Sweat the Small Stuff at Work*, Hodder and Stoughton.

Carr, Allen (1999) *Allen Carr's Easy Way to Stop Smoking*, Penguin Books.

Childre, Doc and Cryer, Bruce (1998) *Freeze Frame: One Minute Stress Management – A Scientifically Proven Technique for Clear Decision Making and Improved Health*, Planetary Publications.

Hay, Louisa L. (1990) *Love Yourself, Heal Your Life Workbook*, Eden Grove Editions.

Hopson, Barrie and Scally, Mike (1993) *Assertiveness: A Positive Process*, Prentice Hall.

Jones, Derek Llewellyn (1993) *Everybody: The Healthy Eating Handbook*, Oxford University Press.

Keeney, Bradford (1997) *The Energy Break*, Newleaf.

Lumsden, Robert (1999) *23 Steps to Success and Achievement*, Thorsons.

O'Brien, Paddy (1993) *Fit to Work*, Sheldon Press.

O'Hara, Valerie (1995) *Wellness at Work: Building Resilience to Job Stress*, New Harbinger Publications.

Schechter, Howard (1995) *Rekindling the Spirit in Work*, Barrytown.

Warren, Deborah Marshall (1999) *Mind Detox*, Thorsons.

Warren, Eve and Toll, Caroline (1994) *The Stress Workbook*, Nicholas Brealey.

BIBLIOGRAPHY

Benson, Herbert (1996) *Timeless Healing*, Scribner.

Bolger, N. (1990) 'Coping as a personality process – a prospective study', *Journal of Personality and Social Psychology*, 59 (3).

Bourne, Edmund J. (1997) *The Anxiety and Phobia Workbook*, New Harbinger Publications.

David, Martha, McKay, Andrew and Eshelman, Elizabeth Robbins (1980) *The Relaxation and Stress Reduction Workbook*, New Harbinger Publications.

Florian, V., Mikulincer, M. and Taubman, O. (1995) 'Does hardiness contribute to mental health during a stressful real life situation? The roles of appraisal and coping', *Journal of Personality and Social Psychology*, 69 (4).

Holland, Jimmie C. and Lewis, Sheldon (1993) 'Emotions and cancer – what do we really know?' in Goleman, Daniel and Gurin, Joel, *Mind Body Medicine*, Consumer Reports.

Loughary, John and Ripley, Theresa (1987) *Dual Career Couples*, United Learning Corporation.

Park, C., Cohen, L.H. and Herb, L. (1990) 'Intrinsic religiousness and religious coping as life stress moderators for Catholics versus Protestants', *Journal of Personality and Social Psychology*, 59 (3).

Spiegel, David (1993) 'Social support – how friends, family and groups can help', in Goleman, Daniel and Gurin, Joel, *Mind Body Medicine*, Consumer Reports.

Turk, Dennis C. and Nash, Justin M. (1993) 'Chronic pain: new ways to cope', in Goleman, Daniel and Gurin, Joel, *Mind Body Medicine*, Consumer Reports.

Williams, Redford B. (1993) 'Hostility and the heart', in Goleman, Daniel and Gurin, Joel, *Mind Body Medicine*, Consumer Reports.

INDEX